Student Solutions Manual
for
Yoshiwara and Yoshiwara's

Prealgebra
Second Edition

Student Solutions Manual
for
Yoshiwara and Yoshiwara's

Prealgebra
Second Edition

Heather R. Foes

THOMSON

™

BROOKS/COLE

Australia • Canada • Mexico • Singapore • Spain • United Kingdom • United States

Printed in the United States of America.
  4   5   6   7   07 06 05

Printer: **Globus Printing & Packaging**
         Minster, Ohio
0-534-38603-2

For more information about our products,
contact us at:
Thomson Learning Academic Resource Center
1-800-423-0563

For permission to use material from this text,
contact us by:
Phone: 1-800-730-2214
Fax: 1-800-730-2215
Web: http://www.thomsonrights.com

Asia
Thomson Learning
5 Shenton Way #01-01
UIC Building
Singapore 068808

Australia/New Zealand
Thomson Learning
102 Dodds Street
Southbank, Victoria 3006
Australia

Canada
Nelson
1120 Birchmount Road
Toronto, Ontario M1K 5G4
Canada

Europe/Middle East/South Africa
Thomson Learning
High Holborn House
50/51 Bedford Row
London WC1R 4LR
United Kingdom

Latin America
Thomson Learning
Seneca, 53
Colonia Polanco
11560 Mexico D.F.
Mexico

Spain/Portugal
Paraninfo
Calle/Magallanes, 25
28015 Madrid, Spain

This book is dedicated to Abby Grace who changed my life forever. I would like to thank Elizabeth Carrico for her thorough checking of these solutions and Rachael Sturgeon for her constant help.

Heather R. Foes

Table of Contents

Student Solutions Manual
for
Yoshiwara and Yoshiwara's

Prealgebra
Second Edition

Homework 1.1A

1. Shirley's laps = Janet's laps - 8

3. Number of batteries = number of packages $\times$ 6

5. Number of pages left = 200 - number of pages read

7. Wages = hours worked $\times$ 7

9. $16.9 + 1.4 = 18.3$ 11. $9.6 - 0.9 = 8.7$ 13. $2.5 \times 18 = 45$

15. $18 \div 24 = 0.75$ 17. $11.2 \div 1.4 = 8$

Homework 1.1B

1. a. Percent of children born outside marriage
    b. 15%
    c. Ireland
    d. Denmark
    e. Greece

3. a. Average temperature in Chicago
    b. 54°F
    c. May
    d. April
    e. November

5. a.

Cause	Falls	Choking on Food	Suffocation or Strangulation	Drowning	Home Fires	Motor Vehicles
Number of Deaths	60	125	350	450	750	1000

    b. Motor vehicles   c. $\dfrac{450}{3400} = 0.132$

    d. Number of deaths due to causes in graph:
        $60 + 125 + 350 + 450 + 750 + 1000 = 2735$
        Number of deaths due to causes not shown in graph:
        $3400 - 2735 = 665$

7. a. U.S.; Mexico
    b. Japan; Mexico
    c. Mexico; by $74 - 33 = 41$ years
    d.
        gap in 1900: $47 - 33 = 14$
        gap now: $80 - 74 = 6$
        The gap has decreased by 8 years.

9. variable: A variable is a letter that stands for a numerical quantity.
    constant: A constant is a number that does not change.
    bar graph: A bar graph uses bars to visually illustrate the values of a variable in different situations.

11.   25% = 0.25   13.   62% = 0.62   15.   6% = 0.06

17.   150% = 1.5   19.   4.6% = 0.046

You can convert a percent to a decimal fraction by moving the decimal point two places to the left.

21.   60% = 0.6   23.   9% = 0.09   25.   8.5% = 0.085

27.   250% = 2.5   29.   0.4% = 0.004

Homework 1.2A

1. a.

Regular price	18	25	54	76	115	130
(Calculation)	18-12	25-12	54-12	76-12	115-12	130-12
Delbert's price	6	13	42	64	103	118

b. You find Delbert's price by subtracting $12 from the regular price.

c. Delbert's price = Regular price − 12

d. Delbert's price = $p - 12$

e.

p	18	25	54	76	115	130
p-12	6	13	42	64	103	118

3. a.

Total calories	400	550	700	925	1200	1500
(Calculation)	400÷5	550÷5	700÷5	925÷5	1200÷5	1500÷5
Calories from fat	80	110	140	185	240	300

b. You find the number of calories from fat allowed by dividing the total calories by 5.

c. Calories from fat = Total calories ÷ 5

d. Calories from fat = $c \div 5$

e.

c	400	550	700	925	1200	1500
c÷5	80	110	140	185	240	300

5. a.

Total bill	20	26	32	48	52	60
(Calculation)	0.15×20	0.15×26	0.15×32	0.15×48	0.15×52	0.15×60
Tip	3.00	3.90	4.80	7.20	7.80	9.00

b. You find the tip by multiplying the total bill by 0.15.

c. Tip = 0.15 × Total bill

d. Tip = $0.15 \times B$

e.

B	20	26	32	48	52	60
0.15×B	3.00	3.90	4.80	7.20	7.80	9.00

7. a. 62%
    b. 50%
    c. S.E. Conference
    d. S.E. Conference
    e. U of Michigan

9. a.

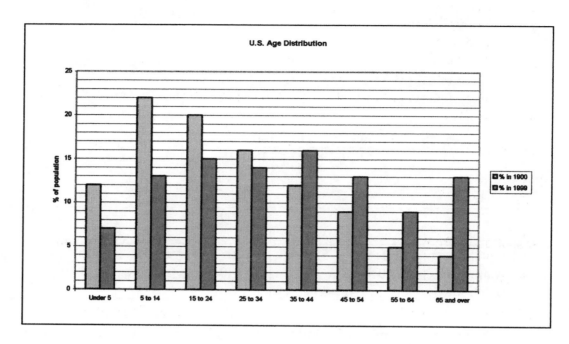

b. 5 to 14; 35 to 44

c. $12 + 22 + 20 = 54\%$
    $7 + 13 + 15 = 35\%$

d. $9 + 5 + 4 = 18\%$
    $13 + 9 + 13 = 35\%$

11. $\dfrac{1}{4} \times 48 = 48 \div 4 = 12$

13. $\dfrac{1}{2} \times 18 = 18 \div 2 = 9$

15. $\dfrac{1}{8} \times 40 = 40 \div 8 = 5$

17. $\dfrac{1}{6} \times 42 = 42 \div 6 = 7$

19. $\dfrac{1}{12} \times 72 = 72 \div 12 = 6$

21. $\dfrac{1}{3} \times 234 = 234 \div 3 = 78$

23. $\dfrac{1}{5} \times 685 = 685 \div 5 = 137$

25. $\dfrac{1}{9} \times 243 = 243 \div 9 = 27$

27. $\dfrac{1}{15} \times 240 = 240 \div 15 = 16$

29. $\dfrac{1}{16} \times 208 = 208 \div 16 = 13$

Homework 1.2 B

1. a. multiplication
    b. subtraction
    c. addition
    d. division
    e. multiplication
    f. addition
    g. division

3. Multiplication can be indicated using a dot or parentheses. For example, 4×5 can be written as $4 \cdot 5$ or $4(5)$.

5. terms; sum

7. $8 + z$

9. $h - 5$

11. $\dfrac{t}{20}$

13. $\dfrac{16}{v}$

15. a subtracted from 7.1

17. the product of 15 and j

19. the quotient of A and c

21. h added to y

23. 2.5 subtracted from H

25. three-fifths of b

27. a. Amount Francine saves $= \dfrac{m}{5}$

    b.

Spending money (dollars)	m	20	25	60	80
Amount saved (dollars)	$\dfrac{m}{5}$	4	5	12	16

29. a. Height $= \dfrac{W}{4}$

    b.

Width of photo (inches)	W	6	10	24	30
Height of photo (inches)	$\dfrac{W}{4}$	$1\dfrac{1}{2}$	$2\dfrac{1}{2}$	6	$7\dfrac{1}{2}$

31. $\dfrac{2}{3}(18) = 2\left(\dfrac{18}{3}\right) = 2(6) = 12$

33. $\dfrac{5}{2}(28) = 5\left(\dfrac{28}{2}\right) = 5(14) = 70$

35. $\dfrac{3}{8}(24) = 3\left(\dfrac{24}{8}\right) = 3(3) = 9$

37. $\dfrac{7}{3}(21) = 7\left(\dfrac{21}{3}\right) = 7(7) = 49$

39. $\dfrac{7}{9}(36) = 7\left(\dfrac{36}{9}\right) = 7(4) = 28$

41. $\dfrac{5}{7}(28) = 5\left(\dfrac{28}{7}\right) = 5 \times 28 \div 7 = 20$

43. $\dfrac{5}{3}(18) = 5\left(\dfrac{18}{3}\right) = 5 \times 18 \div 3 = 30$

45. $\dfrac{4}{5}(16) = 4\left(\dfrac{16}{5}\right) = 4 \times 16 \div 5 = 12.8$

47. $\dfrac{3}{8}(36) = 3\left(\dfrac{36}{8}\right) = 3 \times 36 \div 8 = 13.5$

49. $\dfrac{9}{4}(13) = 9\left(\dfrac{13}{4}\right) = 9 \times 13 \div 4 = 29.25$

Homework 1.3 A

1. $v - 8$

3. $H + 14$

5. $\dfrac{t}{5}$

7. $\dfrac{2}{3}w$

9. $\dfrac{15}{b}$

11. $R - 2.6$

13. $\dfrac{P}{3}$

15. $0.18N$

17. $r - d$

19. $P + I$

21. The difference of x and 3, 3 less than x, x minus 3, x decreased by 3

23. The product of 9 and h, 9 times h, h multiplied by 9, 9 multiplied by h

25. a. 24 divided by 6; $24 \div 6 = 4$

    b. 6 divided by 24; $6 \div 24 = \dfrac{6}{24} = \dfrac{1}{4}$

    c. The ratio of 6 to 24; $6 \div 24 = \dfrac{6}{24} = \dfrac{1}{4}$

    d. The ratio of 24 to 6; $24 \div 6 = 4$

    (a) and (d) are the same.   (b) and (c) are the same.

27. a. Votes needed $= \dfrac{2}{3}p$

    b.

Members present	p	90	96	120	129
Votes needed	$\dfrac{2}{3}p$	60	64	80	86

29.

$3 = \dfrac{3}{4}(4)$ $15 = \dfrac{3}{4}(20)$

$6 = \dfrac{3}{4}(8)$ $75 = \dfrac{3}{4}(100)$

Sale price $= \dfrac{3}{4}p$

31. 70% of 30 = 0.70(30) = 21

33. 15% of 60 = 0.15(60) = 9

35. 32% of 58 = 0.32(58) = 18.56

37. 7% of 83 = 0.07(83) = 5.81

39. 2% of 18 = 0.02(18) = 0.36

Homework 1.3 B

1. a. 1200 − 700 = 500 calories
    b. Subtract c from 1200.
    c.

c	200	350	425	515	640	870
$1200 - c$	1000	850	775	685	560	330

3. a. $\dfrac{600}{60} = 10$ hours
    b. Divide 600 by r.
    c.

r	30	40	45	50	80
$\dfrac{600}{r}$	20	15	$13.\overline{3}$	12	7.5

5. a. 0.10(15) = 1.5 quarts
    b. Multiply q by 0.10.
    c.

q	10	20	45	58	260	1250
$0.10q$	1	2	4.5	5.8	26	125

7. a. $2000(0.07) = $140
    b. Multiply 2000 by r.
    c.

r	6%	6.5%	8%	9.25%	10%
$2000r$	120	130	160	185	200

9. $z - y = 8 - 3 = 5$

11. $9.4 + x = 9.4 + 6 = 15.4$

13. $5x = 5(6) = 30$

15. $xz = 6(8) = 48$

17. $0.4y = 0.4(3) = 1.2$

19. $\dfrac{12}{x} = \dfrac{12}{6} = 2$

21. $\dfrac{y}{9} = \dfrac{3}{9} = \dfrac{1}{3}$

23. $\dfrac{2.4}{x} = \dfrac{2.4}{6} = 0.4$

25. $z - 0.8 = 8 - 0.8 = 7.2$

27.

Length of regular skirt	24	26.5	31	34	L
Length of petite skirt	21.5	24	28.5	31.5	$L-2.5$

29.

Area of room	100	150	200	360	A
Cost of carpet	800	1200	1600	2880	$8A$

31. $\dfrac{1}{5}(75) = \dfrac{75}{5} = 15$ 33. $0.3(20) = 6$

35. $\dfrac{3}{4}(24) = 3\left(\dfrac{24}{4}\right) = 3(6) = 18$ 37. $80\% \text{ of } 60 = 0.80(60) = 48$

39. $0.8(74) = 59.2$ 41. $\dfrac{1}{4}(39) = 9.75$

43. $43\% \text{ of } 17 = 0.43(17) = 7.31$ 45. $\dfrac{3}{8}(95) = 35.625$

Homework 1.4 A

1. $m - 15$

3. $m + 15$

5. $15 - m$

7. $12p$

9. $\dfrac{p}{12}$

11. $\dfrac{12}{p}$

13. $\dfrac{36}{c}$

15. $c + 36$

17. $36 - c$

19. $36c$

21.   Step 1: The cost of a water filter is unknown.
    Step 2: Cost of a water filter: w
    Step 3: "times" suggests multiplication: $3w$

23.   Step 1: The radius of the circle is unknown.
    Step 2: Radius of the circle: r
    Step 3: "decreased by" suggests subtraction: $r - 6$

25.   Step 1: The checking account balance is unknown.
    Step 2: Checking account balance: B
    Step 3: "of" suggests multiplication: $\dfrac{2}{5}B$

27.   Step 1: The price of the pizza is unknown.
    Step 2: Price of the pizza: p
    Step 3: "divided by" suggests division: $\dfrac{p}{4}$

29.   Step 1: The perimeter of the triangle is unknown.
    Step 2: Perimeter of the triangle: P
    Step 3: "diminished" suggests subtraction: $P - 15$

31. $\dfrac{1}{4} = 0.25$

33. $\dfrac{5}{8} = 0.625$

35. $\dfrac{2}{5} = 0.4$

37. $\dfrac{7}{20} = 0.35$

39. $\dfrac{5}{16} = 0.3125$

41. $6\dfrac{1}{4} = 1 \div 4 + 6 = 6.25$

43. $4\dfrac{1}{8} = 1 \div 8 + 4 = 4.125$ 45. $7\dfrac{2}{5} = 2 \div 5 + 7 = 7.4$

47. $15\dfrac{7}{16} = 7 \div 16 + 15 = 15.4375$ 49. $2\dfrac{9}{20} = 9 \div 20 + 2 = 2.45$

Homework 1.4 B

1. a. distance $= rt = 150t$
   b. If $t = 3$, the distance is $150(3) = 450$ miles.
      If $t = 8$, the distance is $150(8) = 1200$ miles.
      If $t = 4.5$, the distance is $150(4.5) = 675$ miles.

3. a. time $= \dfrac{d}{r} = \dfrac{1000}{r}$
   b. If $r = 50$, the time is $\dfrac{1000}{50} = 20$ hours.
      If $r = 400$, the time is $\dfrac{1000}{400} = 2.5$ hours.
      If $r = 8$, the time is $\dfrac{1000}{8} = 125$ hours.

5. a. Interest $= Prt = 1600(0.04)t$
   b. If $t = 1$, the interest is $1600(0.04)(1) = \$64$.
      If $t = 2$, the interest is $1600(0.04)(2) = \$128$.
      If $t = 5$, the interest is $1600(0.04)(5) = \$320$.

7. a. Interest $= Prt = 6000(3)r$
   b. If $r = 0.05$, the interest is $6000(3)(0.05) = \$900$.
      If $r = 0.0825$, the interest is $6000(3)(0.0825) = \$1485$.
      If $r = 0.12$, the interest is $6000(3)(0.12) = \$2160$.

9. a. Let p stand for the amount of fruit punch.
      Amount of grape juice $= 0.20p$
   b. If $p = 3$, the amount of grape juice is $0.20(3) = 0.6$ quarts.
      If $p = 14$, the amount of grape juice is $0.20(14) = 2.8$ quarts.
      If $p = 60$, the amount of grape juice is $0.20(60) = 12$ quarts.

11. a. Laureen's quiz average $= \dfrac{S}{12}$
    b. If S $= 96$, Laureen's quiz average is $\dfrac{S}{12} = \dfrac{96}{12} = 8$.
       If S $= 78$, Laureen's quiz average is $\dfrac{S}{12} = \dfrac{78}{12} = 6.5$.
       If S $= 111$, Laureen's quiz average is $\dfrac{S}{12} = \dfrac{111}{12} = 9.25$.

13.   a.   Let F stand for the amount of flour Amanda has.

Amount of flour required $= F + 3$

     b.   If $F = 8$, the amount of flour required is $8 + 3 = 11$ cups.

15.   a.   Let T stand for Jamal's total points.

Jamal's homework grade $= \dfrac{T}{600}$

     b.   If $T = 480$, Jamal's homework grade is $\dfrac{480}{600} = 0.80$.

17.   a. and b.   For Singapore, time $= \dfrac{d}{r} = \dfrac{10}{37} \approx 0.270$ hours.

For Sydney, time $= \dfrac{d}{r} = \dfrac{10}{11} \approx 0.909$ hours.

City	Rush Hour Traffic Flow (mph)	Commute Time (Hours)
Singapore	37	0.270
Sydney	11	0.909
Hong Kong	12	0.833
Bangkok	13	0.769
Shanghai	15	0.667
Manila	7	1.429
Jakarta	16	0.625
Los Angeles	19	0.526
Seoul	14	0.714
Tokyo-Yokohama	28	0.357

   c.   Manila; Seoul   d.   Hong Kong, $\dfrac{4}{5}(15) = 12$

Tokyo-Yokohama, $1.75(16) = 28$

19. $\dfrac{2}{3} = 0.\overline{6}$ 21. $\dfrac{5}{6} = 0.8\overline{3}$ 23. $\dfrac{7}{9} = 0.\overline{7}$

25. $\dfrac{8}{11} = 0.\overline{72}$ 27. $\dfrac{1}{12} = 0.08\overline{3}$

Homework 1.5

1.

Simone's age	6	10	15	**18**	30	**24**	s
Rachel's age	14	**18**	**23**	26	**38**	32	$s+8$

    a.   Add 8 to Simone's age.
    b.   Subtract 8 from Rachel's age.
    c. $r = s + 8$
    d.   Add 8 to s.
    e.   Subtract 8 from r.

3.

Bank statement	100	138	188	**276**	332	**352**	b
Actual balance	24	**62**	**112**	200	**256**	276	**$b-76$**

    a.   Subtract 76 from the bank statement.
    b.   Add 76 to the actual balance.
    c. $a = b - 76$
    d.   Subtract 76 from b.
    e.   Add 76 to a.

5.   a. q is 8 less than r.   7.   a. m is h divided by 3.

      b. $q = r - 8$ b. $m = \dfrac{h}{3}$

9.   a. k is 0.6 more than c.   11.   a d is $\dfrac{3}{4}$ of a.
      b. $k = c + 0.6$

                                                  b. $d = \dfrac{3}{4}a$

13. $8p = 40$ 15. $p - 8 = 40$ 17. $p + 8 = 40$

19. $8p = 40$ 21. $b - 18 = 10$ 23. $18 - b = 10$

25. $18 - b = 10$ 27. $16t = 80$ 29. $16t = 80$

31. $n - 5 = 30$ 33. $5n = 30$ 35. $n + 5 = 30$

37. $5n = 30$

Fraction	Decimal	Percent	Fraction	Decimal	Percent
$\frac{1}{2}$	0.5	50%	$\frac{3}{5}$	0.6	60%
$\frac{1}{3}$	$0.\overline{3}$	$33\frac{1}{3}\%$	$\frac{4}{5}$	0.8	80%
$\frac{2}{3}$	$0.\overline{6}$	$66\frac{2}{3}\%$	$\frac{1}{8}$	0.125	12.5%
$\frac{1}{4}$	0.25	25%	$\frac{3}{8}$	0.375	37.5%
$\frac{3}{4}$	0.75	75%	$\frac{5}{8}$	0.625	62.5%
$\frac{1}{5}$	0.2	20%	$\frac{7}{8}$	0.875	87.5%
$\frac{2}{5}$	0.4	40%	1	1.0	100%

39. $0.125 = \frac{1}{8}$

41. $\frac{3}{8} = 0.375$

43. $0.\overline{6} = \frac{2}{3}$

45. $\frac{2}{5} = 0.4$

47. $\frac{1}{3} = 0.\overline{3}$

49. $0.625 = \frac{5}{8}$

Chapter 1 Review

1. product

3. factors

5. ratio

7. variable

9. evaluate

11. terms

13. a. To find the number of years Chelsea has studied dance, add 3 to the number of years Brianna has studied dance.

    b. Let B stand for the number of years Brianna has studied dance.
    Number of years Chelsea has studied dance $= B + 3$

15. a. To find the amount Vanda has left for living expenses, subtract $500 from her paycheck.

    b. Let p stand for the amount of Vanda's paycheck.
    Amount left for living expenses $= p - 500$

17. a. July 4; 95°F
    b. July 9; 71°F
    c. July 7; 5°F
    d. July 8; 36°F

19. $1.5a = 1.5(2.8) = 4.2$

21. $\dfrac{6.4}{c} = \dfrac{6.4}{1.6} = 4$

23. $2pq = 2(3)(14) = 84$

25. distance $= rt$; $r =$ rate, $t =$ time

27. interest $= Prt$; $P =$ principal, $r =$ interest rate, $t =$ time

29. percentage $= rW$
    percentage $= 0.22(9,400,000) = 2,068,000$ square miles

31. time $= \dfrac{d}{r}$

    time $= \dfrac{12}{8} = 1.5$ hours

33. a. the fraction of a bottle in each glass
    b. $50n = 10$

35. a. the amount of the tax increase
    b. $\dfrac{n}{50} = 10$

37. $v = d + 0.4$

39. $y = x + 2.5$

41. $130\% = 1.3$

43. $29\% = 0.29$

45. $\dfrac{1}{3}$ of $36 = \dfrac{1}{3}(36) = \dfrac{36}{3} = 12$

47. $\dfrac{1}{8}$ of $48 = \dfrac{1}{8}(48) = \dfrac{48}{8} = 6$

49. $\dfrac{2}{5}(25) = 2\left(\dfrac{25}{5}\right) = 2(5) = 10$

51. $\dfrac{7}{4}(20) = 7\left(\dfrac{20}{4}\right) = 7(5) = 35$

53. 125% of $60 = 1.25(60) = 75$

55. 6% of $39 = 0.06(39) = 2.34$

57. $5\dfrac{3}{4} = 5 + \dfrac{3}{4} = 5 + 0.75 = 5.75$

59. $7\dfrac{3}{5} = 7 + \dfrac{3}{5} = 7 + 0.6 = 7.6$

61. $\dfrac{2}{11} = 0.\overline{18}$

63. $4\dfrac{1}{6} = 4.1\overline{6}$

Homework 2.1 A

1.   An equation is a statement that two expressions are equal.

3.   Equation; There are two expressions and an equal sign between them.

5.   Equation; There are two expressions and an equal sign between them.

7.   Equation; There are two expressions and an equal sign between them.

9.   Equation; There are two expressions and an equal sign between them.

11.   Not an equation; There is no equal sign.

13.   Equation; There are two expressions and an equal sign between them.

15.   Equation; There are two expressions and an equal sign between them.

17.
$$x + 19 = 36$$
$$17 + 19 = 36?$$
$$36 = 36 \text{ (True)}$$
The given value is a solution of the equation.

19.
$$t - 7 = 25$$
$$18 - 7 = 25?$$
$$11 = 25 \text{ (False)}$$
The given value is not a solution of the equation.

21.
$$12v = 60$$
$$12(5) = 60?$$
$$60 = 60 \text{ (True)}$$
The given value is a solution of the equation.

23.
$$\frac{z}{15} = 3$$
$$\frac{5}{15} = 3?$$
$$\frac{1}{3} = 3 \text{ (False)}$$
The given value is not a solution of the equation.

25.
$$3.4 - q = 0.8$$
$$3.4 - 1.6 = 0.8?$$
$$1.8 = 0.8 \text{ (False)}$$
The given value is not a solution of the equation.

27.
$$5n = 28.2$$
$$5(23.2) = 28.2?$$
$$116 = 28.2 \text{ (False)}$$
The given value is not a solution of the equation.

29.   a.

x	0.6	0.8	1.0	1.2	1.4	1.6
$1.6x$	0.96	1.28	1.6	1.92	2.24	2.56

b.
$$1.6x = 1.92$$
$$x = 1.2$$

31. a.

z	14	21	28	35	42	49
$87-z$	73	66	59	52	45	38

b.

$$87 - z = 38$$
$$z = 49$$

33. a. $w - 4.5$
     b. $n = w - 4.5$
     c.

w	20	**18**	15
n	**15.5**	13.5	**10.5**

35. a. $1.60p$
     b. $d = 1.60p$
     c.

p	5	**10**	20
d	**8**	16	**32**

37. $8\frac{1}{4}\% = 8.25\% = 0.0825$

39. $108\frac{3}{5}\% = 108.6\% = 1.086$

41. $\frac{5}{8}\% = 0.625\% = 0.00625$

43. $33\frac{1}{3}\% = 33.\overline{3}\% = 0.\overline{3}$

45. $10\frac{3}{8}\% = 10.375\% = 0.10375$

Homework 2.1B

1.   When you solve an equation, you find the solution(s) of the equation.

3.   *Step 1*: Determine which operation has been performed on the variable.
     Step 2: Perform the opposite operation on both sides of the equation.
     Step 3: Check the solution.

5.

$$
\begin{array}{ll}
b + 8 = 15 & \text{Check: } b + 8 = 15 \\
\underline{-8 \quad -8} & \qquad\quad 7 + 8 = 15? \\
\quad b = 7 & \qquad\qquad 15 = 15
\end{array}
$$

7.

$$
\begin{array}{ll}
3 + t = 5 & \text{Check: } 3 + t = 5 \\
\underline{-3 \quad -3} & \qquad\quad 3 + 2 = 5? \\
\quad t = 2 & \qquad\qquad 5 = 5
\end{array}
$$

9.

$$
\begin{array}{ll}
4 = x + 1 & \text{Check: } 4 = x + 1 \\
\underline{-1 \quad -1} & \qquad\quad 4 = 3 + 1? \\
\quad 3 = x & \qquad\qquad 4 = 4 \\
\quad x = 3 &
\end{array}
$$

11.

$$
\begin{array}{ll}
h + 3.9 = 6.8 & \text{Check: } h + 3.9 = 6.8 \\
\underline{-3.9 \quad -3.9} & \qquad 2.9 + 3.9 = 6.8? \\
\quad h = 2.9 & \qquad\qquad 6.8 = 6.8
\end{array}
$$

13.

$$
\begin{array}{ll}
w - 7 = 5 & \text{Check: } w - 7 = 5 \\
\underline{+7 \quad +7} & \qquad\quad 12 - 7 = 5? \\
\quad w = 12 & \qquad\qquad 5 = 5
\end{array}
$$

15.

$$
\begin{array}{ll}
16 = d - 8 & \text{Check: } 16 = d - 8 \\
\underline{+8 \qquad +8} & \qquad\quad 16 = 24 - 8? \\
24 = d & \qquad\qquad 16 = 16
\end{array}
$$

17.

$$p - 15 = 15 \qquad \text{Check: } p - 15 = 15$$
$$\underline{+15 \quad +15} \qquad\qquad 30 - 15 = 15?$$
$$p = 30 \qquad\qquad\qquad 15 = 15$$

19.

$$x - 0.05 = 0.8 \qquad \text{Check: } x - 0.05 = 0.8$$
$$\underline{+0.05 \quad +0.05} \qquad\qquad 0.85 - 0.05 = 0.8?$$
$$x = 0.85 \qquad\qquad\qquad 0.8 = 0.8$$

21.

$$5g = 30 \qquad \text{Check: } 5g = 30$$
$$\frac{5g}{5} = \frac{30}{5} \qquad\qquad 5(6) = 30?$$
$$g = 6 \qquad\qquad\qquad 30 = 30$$

23.

$$78 = 6h \qquad \text{Check: } 78 = 6h$$
$$\frac{78}{6} = \frac{6h}{6} \qquad\qquad 78 = 6(13)?$$
$$13 = h \qquad\qquad\qquad 78 = 78$$
$$h = 13$$

25.

$$15p = 75 \qquad \text{Check: } 15p = 75$$
$$\frac{15p}{15} = \frac{75}{15} \qquad\qquad 15(5) = 75?$$
$$p = 5 \qquad\qquad\qquad 75 = 75$$

27.

$$2z = 9.6 \qquad \text{Check: } 2z = 9.6$$
$$\frac{2z}{2} = \frac{9.6}{2} \qquad\qquad 2(4.8) = 9.6?$$
$$z = 4.8 \qquad\qquad\qquad 9.6 = 9.6$$

29.

$$\frac{m}{4} = 7 \qquad \text{Check}: \frac{m}{4} = 7$$
$$4\left(\frac{m}{4}\right) = (7)4 \qquad\qquad \frac{28}{4} = 7$$
$$m = 28 \qquad\qquad\qquad 7 = 7$$

31.

$$\frac{x}{3} = 14 \qquad \text{Check}: \frac{x}{3} = 14$$
$$3\left(\frac{x}{3}\right) = (14)3 \qquad\qquad \frac{42}{3} = 14?$$
$$x = 42 \qquad\qquad\qquad 14 = 14$$

33.

$$13 = \frac{s}{16} \qquad \text{Check}: 13 = \frac{s}{16}$$
$$16(13) = \left(\frac{s}{16}\right)16 \qquad\qquad 13 = \frac{208}{16}?$$
$$208 = s \qquad\qquad\qquad 13 = 13$$
$$s = 208$$

35.

$$\frac{v}{6} = 1.3 \qquad \text{Check}: \frac{v}{6} = 1.3$$
$$6\left(\frac{v}{6}\right) = (1.3)6 \qquad\qquad \frac{7.8}{6} = 1.3?$$
$$v = 7.8 \qquad\qquad\qquad 1.3 = 1.3$$

37.

$$d - 16 = 16 \qquad \text{Check: } d - 16 = 16$$
$$\underline{+16 \quad +16} \qquad\qquad 32 - 16 = 16?$$
$$d = 32 \qquad\qquad\qquad 16 = 16$$

39.

$$0 = f - 37 \qquad \text{Check: } 0 = f - 37$$
$$\underline{+37 \qquad +37} \qquad\qquad 0 = 37 - 37?$$
$$37 = f \qquad\qquad\qquad 0 = 0$$
$$f = 37$$

41.

$$\frac{x}{5} = 0 \qquad \text{Check}: \frac{x}{5} = 0$$
$$5\left(\frac{x}{5}\right) = (0)5 \qquad\qquad \frac{0}{5} = 0?$$
$$x = 0 \qquad\qquad\qquad 0 = 0$$

43.

$$47.3 = 24.8 + v \qquad \text{Check:}$$
$$\underline{-24.8 \quad -24.8} \qquad\qquad 47.3 = 24.8 + v$$
$$22.5 = v \qquad\qquad\qquad 47.3 = 24.8 + 22.5?$$
$$v = 22.5 \qquad\qquad\qquad 47.3 = 47.3$$

45.

$$68.89 = 8.3n \qquad \text{Check:}$$

$$\frac{68.89}{8.3} = \frac{8.3n}{8.3} \qquad 68.89 = 8.3n?$$

$$8.3 = n \qquad 68.89 = 8.3(8.3)$$

$$n = 8.3 \qquad 68.89 = 68.89$$

47.

$$\frac{w}{38} = 1.1 \qquad \text{Check}: \frac{w}{38} = 1.1$$

$$38\left(\frac{w}{38}\right) = (1.1)38 \qquad \frac{41.8}{38} = 1.1?$$

$$w = 41.8 \qquad 1.1 = 1.1$$

49. 8.3 is larger.

51. 4.0800 is larger

53. 0.20 is larger.

55. 606.060 is larger.

57. 0.0110 is larger.

Homework 2.2 A

1. The perimeter of a figure measures the distance around the border of the figure. Two typical units for perimeter are inches and feet.

3. Perimeter: $5 + 5 + 5 + 5 = 20$ cm
   Area: 25 square cm

5. Perimeter: $4 + 9 + 4 + 9 = 26$ cm
   Area: 36 square cm

7. Perimeter:
   $3 + 1 + 2 + 2 + 1 + 3 = 12$ cm
   Area: 5 square cm

9. Perimeter: $4 + 1 + 1 + 1 + 1 + 1 + 1 + 1 + 1 + 4 = 16$ cm
   Area: 10 square cm

11.
$$59 = 32 + m$$
$$\underline{-32 \quad -32}$$
$$27 = m$$
$$m = 27$$
Check: $59 = 32 + m$
$59 = 32 + 27?$
$59 = 59$

13.
$$12.3 = a - 9.1$$
$$\underline{+9.1 \qquad +9.1}$$
$$21.4 = a$$
$$a = 21.4$$
Check:
$12.3 = a - 9.1$
$12.3 = 21.4 - 9.1?$
$12.3 = 12.3$

15.
$$4.2d = 33.6$$
$$\frac{4.2d}{4.2} = \frac{33.6}{4.2}$$
$$d = 8$$
Check: $4.2d = 33.6$
$4.2(8) = 33.6?$
$33.6 = 33.6$

17.
$$\frac{t}{0.9} = 0.2$$
$$0.9\left(\frac{t}{0.9}\right) = (0.2)0.9$$
$$t = 0.18$$
Check : $\frac{t}{0.9} = 0.2$
$\frac{0.18}{0.9} = 0.2?$
$0.2 = 0.2$

19. a. Let L stand for the length of the assignment.
   $$\frac{2}{3}L$$
   b. $\frac{2}{3}(78) = 52$ pages

21.
$$\frac{1}{8} = 0.125$$
$\frac{1}{8}$ is larger.

23.
$$\frac{3}{8} = 0.375$$
0.38 is larger.

25.
$$\frac{1}{3} = 0.\overline{3}$$
$\frac{1}{3}$ is larger.

27.
$$\frac{2}{9} = 0.\overline{2}$$
0.23 is larger.

29.
$$\frac{3}{11} = 0.\overline{27}$$
$\frac{3}{11}$ is larger.

Homework 2.2 B

1.   a. $3\frac{1}{2}$ inches

      b.   8.9 centimeters

3.   a. $2\frac{1}{2}$ inches

      b.   6.4 centimeters

5.   a.   Answers may vary.

      b.   Answers may vary.

7.   Since there are 10 mm in each cm, multiply by 10.

      a.   3 cm = <u>30</u> mm

      b.   1.8 cm = <u>18</u> mm

      c.   0.6 cm = <u>6</u> mm

      d.   0.25 cm = <u>2.5</u> mm

9.   a. $\frac{1}{2}$ in $= \frac{1\cdot 4}{2\cdot 4}$ in $= \frac{\mathbf{4}}{8}$ in

      b. $\frac{3}{4}$ in $= \frac{3\cdot 4}{4\cdot 4}$ in $= \frac{\mathbf{12}}{16}$ in

      c. $\frac{5}{8}$ in $= \frac{5\cdot 2}{8\cdot 2}$ in $= \frac{\mathbf{10}}{16}$ in

      d. $\frac{5}{4}$ in $= \frac{5\cdot 2}{4\cdot 2}$ in $= \frac{\mathbf{10}}{8}$ in

11.   Area: 18 square cm

      Perimeter: 6 + 6 + 8.5 = 20.5 cm

13.   Area: 22.5 square cm

      Perimeter: 5.4 + 8.6 + 9 = 23 cm

15.   Area: 41 square cm

      Perimeter: 3 + 2.8 + 3 + 2.8 + 3 + 2.8 + 3 + 2.8 = 23.2 cm

17.   a

19.   b

21.   2.4

23.   21.1

25.   8.2

27.   48.8

29.   0.91

Homework 2.3 A

1. A formula is an equation that relates two or more variables.

3. $A = \dfrac{S}{n}$; A = average value, S = sum of values, n = number of values

5. $e = \dfrac{m}{g}$; e = efficiency, m = miles, g = gallons

7. $P = rW$; P = part, r = percentage rate, W = whole

9. $s = p - d$; s = sale price, p = regular price, d = discount

11.
$$d = rt$$
$$d = (14)(6)$$
$$d = 84 \text{ miles}$$

13.
$$P = R - C$$
$$P = 8200 - 6835$$
$$P = \$1365$$

15.
$$s = p - d$$
$$s = 360 - 43.20$$
$$s = \$316.80$$

17.
$$e = \dfrac{m}{g}$$
$$e = \dfrac{180}{24}$$
$$e = 7.5 \text{ miles}$$
$$\text{per gallon}$$

19.
$$u = \dfrac{p}{n}$$
$$u = \dfrac{810}{18}$$
$$u = \$45$$

21.
$$P = rW$$
$$P = 0.35(1500)$$
$$P = \$525$$

23. a. yes

    b. $12 \stackrel{?}{=} 2 + 4 + 6$
       $12 = 12$
       yes

25. a. no; There is no equal sign.

27. a. yes

    b. $7(3) \stackrel{?}{=} 10$
       $21 \neq 10$
       no

29. a. yes

    b. $24 - 9 \stackrel{?}{=} 13 + 2$
       $15 = 15$
       yes

31. a. no; There is nothing on the right side of the equal sign.

33.
a. w is 15 less than t.
b. $w = t - 15$

t	w
40	25
55	40
60	45
80	65
90	**75**
110	95

35.
a. H is 4 times D.
b. $H = 4D$

D	H
3	12
5	20
6	24
9	36
11	**44**
16	64

37.
a. 28.467
b. 28.5

39.
a. 5.667
b. 5.7

41.
a. 0.006
b. 0.0

43.
a. 5.000
b. 5.0

45.
a. 1.000
b. 1.0

Homework 2.3 B

1.
$$0.38 + f = 0.72$$
$$\underline{-0.38 \qquad -0.38}$$
$$f = 0.34$$

3.
$$y - 43 = 13$$
$$\underline{+43 \quad +43}$$
$$y = 56$$

5.
$$10.8 = 0.6u$$
$$\frac{10.8}{0.6} = \frac{0.6u}{0.6}$$
$$18 = u$$
$$u = 18$$

7.
$$36 = \frac{p}{1.5}$$
$$1.5(36) = \left(\frac{p}{1.5}\right)1.5$$
$$54 = p$$
$$p = 54$$

9. a.
$$e = \frac{m}{g}$$
$$26 = \frac{m}{14}$$

b.
$$26 = \frac{m}{14}$$
$$14(26) = \left(\frac{m}{14}\right)14$$
$$364 = m$$
The car must travel 364 miles.

11. a.
$$P = R - C$$
$$1500 = R - 215$$

b.
$$1500 = R - 215$$
$$\underline{+215 \qquad +215}$$
$$1715 = R$$
They will need $1715.
The airliner can fly 4,000 miles without refueling.

13. a.
$$P = rW$$
$$162 = 0.36W$$

   b.
$$162 = 0.36W$$
$$\frac{162}{0.36} = \frac{0.36W}{0.36}$$
$$450 = W$$
There are 450 students in the freshman class.

15. a.
$$I = Prt$$
$$98.80 = 1300r(1)$$

   b.
$$98.80 = 1300r(1)$$
$$98.80 = 1300r$$
$$\frac{98.80}{1300} = \frac{1300r}{1300}$$
$$0.076 = r$$
The certificate earned 7.6% interest.

17. a.
$$P = rW$$
$$540 = 0.45W$$

   b.
$$540 = 0.45W$$
$$\frac{540}{0.45} = \frac{0.45W}{0.45}$$
$$1200 = W$$
1200 people voted.

19. a.
$$I = Prt$$
$$171 = P(0.095)(1)$$

   b.
$$171 = P(0.095)(1)$$
$$171 = P(0.095)$$
$$\frac{171}{0.095} = \frac{P(0.095)}{0.095}$$
$$1800 = P$$
Staci deposited $1800 in the T-bill.

21. Perimeter: $2 + 1 + \frac{1}{2} + 2 + \frac{1}{2} + 1 + 2 + 4 = 13$ cm

   Area: 9 square cm

23. Perimeter: $8 + 5 + 5 + 3 + 5 + 2 = 28$ cm
   Area: 28 square cm

25. $\frac{5}{6} = 0.833\cdots \approx 0.83$

27. $\frac{2}{7} = 0.285\cdots \approx 0.29$

29. $\frac{5}{13} = 0.384\cdots \approx 0.38$

31. $\frac{13}{15} = 0.866\cdots \approx 0.87$

33. $\frac{7}{37} = 0.189\cdots \approx 0.19$

Homework 2.4 A

1.
$$A = \frac{1}{2}bh$$
$$A = \frac{1}{2}(8)(5)$$
$$A = 20 \text{ square units}$$

3.
$$A = bh$$
$$A = (5)(2)$$
$$A = 10 \text{ square units}$$

5.
$$A = \frac{1}{2}bh$$
$$A = \frac{1}{2}(7)(4)$$
$$A = 14 \text{ square units}$$

7.
$$A = lw$$
$$A = (16)(9)$$
$$A = 144 \text{ square ft}$$

9.
$$A = \frac{1}{2}bh$$
$$A = \frac{1}{2}(12)(6)$$
$$A = 36 \text{ square m}$$

11.
$$A = bh$$
$$A = (3)(4)$$
$$A = 12 \text{ square cm}$$

13.
$$I = Prt$$
$$I = (20000)(0.03)(5)$$
$$I = \$3,000$$

15.
$$A = \frac{S}{n}$$
$$S = 500 + 1500 + 4000 = 6000$$
$$n = 2\frac{1}{2}(60) = 2.5(60) = 150$$
$$A = \frac{6000}{150} = 40 \text{ words per minute}$$

17.
$$4 = \frac{2}{3}(6)$$
$$y \text{ is } \frac{2}{3} \text{ of } x$$
$$y = \frac{2}{3}x$$

19.
$$12 = \frac{3}{2}(8)$$
$$R \text{ is } \frac{3}{2} \text{ of } N$$
$$R = \frac{3}{2}N$$

21.
$$\frac{1}{6} \text{ of } 11 = \frac{1}{6} \times 11 = 11 \div 6 = 1.833\cdots \approx 1.83$$

23.
$$\frac{1}{5} \text{ of } 32 = \frac{1}{5} \times 32 = 32 \div 5 = 6.4$$

25.
$$\frac{1}{7} \text{ of } 25 = \frac{1}{7} \times 25 = 25 \div 7 = 3.571\cdots \approx 3.57$$

27.
$$\frac{1}{12} \text{ of } 35 = \frac{1}{12} \times 35 = 35 \div 12 = 2.916\cdots$$
$$\approx 2.92$$

29.
$$\frac{1}{8} \text{ of } 17 = \frac{1}{8} \times 17 = 17 \div 8 = 2.125 \approx 2.13$$

Homework 2.4 B

1.

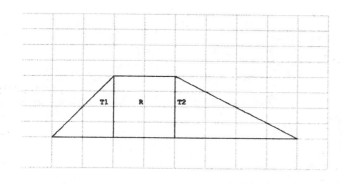

Area of T1: $A = \frac{1}{2}bh = \frac{1}{2}(2)(4) = 4$

Area of R: $A = lw = 2(4) = 8$

Area of T2: $A = \frac{1}{2}bh = \frac{1}{2}(4)(4) = 8$

Total area $= 4 + 8 + 8 = 20$ square cm

3.

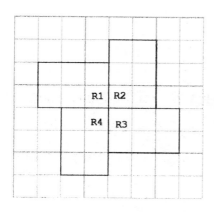

Area of R1: $A = lw = (2)(3) = 6$

Area of R2: $A = lw = (2)(3) = 6$

Area of R3: $A = lw = (2)(3) = 6$

Area of R4: $A = lw = (2)(3) = 6$

Total area $= 6 + 6 + 6 + 6 = 24$ square cm

5.

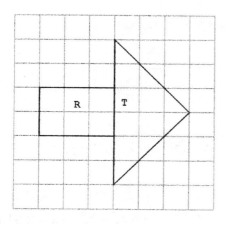

Area of R: $A = lw = (2)(3) = 6$

Area of T: $A = \frac{1}{2}bh = \frac{1}{2}(6)(3) = 9$

Total area $= 6 + 9 = 15$ square cm

7.

$A = \frac{1}{2}bh$

$A = \frac{1}{2}(18)(7)$

$A = 63$ square m

9.

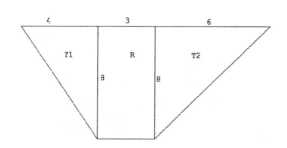

Area of T1: $A = \frac{1}{2}bh = \frac{1}{2}(4)(8) = 16$

Area of R: $A = lw = (3)(8) = 24$

Area of T2: $A = \frac{1}{2}bh = \frac{1}{2}(6)(8) = 24$

Total area $= 16 + 24 + 24 = 64$ square yd

11. $P = 12 + 15 + 12 + 6 + 6 + 3 + 6 + 6 = 66$ m

$A = $ (area of large rectangle) $-$ (area of removed rectangle)

$= (lw) - (lw)$

$= (15 \cdot 12) - (6 \cdot 3)$

$= 180 - 18$

$= 162$ square m

13.

Area $= $ (area of larger triangle) $-$ (area of removed triangle)

$= \left(\frac{1}{2}bh\right) - \left(\frac{1}{2}bh\right)$

$= \left(\frac{1}{2} \cdot 1.7 \cdot 1.7\right) - \left(\frac{1}{2} \cdot 0.7 \cdot 0.7\right)$

$= 1.445 - 0.245$

$= 1.2$ square in.

15.

Area $= $ (area of parallelogram) $-$ (area of removed triangles)

$= (bh) - 2\left(\frac{1}{2}bh\right)$

$= (3 \cdot 2.6) - 2\left(\frac{1}{2} \cdot 1.5 \cdot 1.3\right)$

$= 7.8 - 1.95$

$= 5.85$ square in.

17.

Area $= $ (area of rectangle) $+$ (area of triangles)

$= (lw) - 2\left(\frac{1}{2}bh\right)$

$= (2.8 \cdot 2.5) + 2\left(\frac{1}{2} \cdot 2.4 \cdot 2.5\right)$

$= 7 + 6$

$= 13$ square in.

19.

$$A = (\text{area of square}) + (\text{area of 4 triangles})$$

$$= (lw) + 4\left(\frac{1}{2}bh\right) \qquad h = \frac{12 - 8.5}{2} = 1.75$$

$$= (8.5 \cdot 8.5) + 4\left(\frac{1}{2} \cdot 8.5 \cdot 1.75\right)$$

$$= 72.25 + 29.75$$

$$= 102 \text{ square in.}$$

21.   Red area: 8 square cm
Blue area: 4 square cm
Yellow area: 24 square cm

23.   Yellow area: $\frac{1}{2}bh = \frac{1}{2}(4)(2) = 4$ square cm

Blue area: $\frac{1}{2}bh + \frac{1}{2}bh = \frac{1}{2}(4)(2) + \frac{1}{2}(2)(2) = 4 + 2 = 6$ square cm

Red area = (area of square) − (white area) − (blue area)

$$= 16 - 4 - 6$$

$$= 6 \text{ square cm}$$

25.   a.

$$d = rt$$
$$364 = r(7)$$

b.

$$364 = r(7)$$
$$\frac{364}{7} = \frac{r(7)}{7}$$
$$52 = r$$

She drove 52 miles per hour.

27.   a.

$$d = rt$$
$$2800 = 560t$$

b.

$$2800 = 560t$$
$$\frac{2800}{560} = \frac{560t}{560}$$
$$5 = t$$

The trip takes 5 hours.

29.   a.

$$A = \frac{S}{n}$$
$$13.25 = \frac{S}{16}$$

b.

$$13.25 = \frac{S}{16}$$
$$16(13.25) = \left(\frac{S}{16}\right)16$$
$$212 = S$$

She collected a total of $212.

31. $\frac{2}{7}(20) = 2 \cdot 20 \div 7 = 5.7142 \cdots \approx 5.714$

33. $\frac{7}{6}(25) = 7 \cdot 25 \div 6 = 29.1666 \cdots \approx 29.167$

35. $\frac{4}{9}(38) = 4 \cdot 38 \div 9 = 16.8888 \cdots \approx 16.889$

37. $\frac{5}{3}(8) = 5 \cdot 8 \div 3 = 13.3333 \cdots \approx 13.333$

39. $\frac{11}{12}(37) = 11 \cdot 37 \div 12 = 33.9166 \cdots \approx 33.917$

Homework 2.5 A

1.   a. $6n = 162$

     b.

$$6n = 162$$
$$\frac{6n}{6} = \frac{162}{6}$$
$$n = 27$$

3.   a. $n - 24 = 38$

     b.

$$n - 24 = 38$$
$$\underline{+24 \quad +24}$$
$$n = 62$$

5.   a. $\dfrac{n}{2.5} = 6.6$

     b.

$$\frac{n}{2.5} = 6.6$$
$$2.5\left(\frac{n}{2.5}\right) = (6.6)2.5$$
$$n = 165$$

7.   a. $20 = n + 15.3$

     b.

$$20 = n + 15.3$$
$$\underline{-15.3 \quad -15.3}$$
$$4.7 = n$$
$$n = 4.7$$

9.   a.

Country	Family Size	Average Age at Marriage (Women)
Iraq	7.1	22.3
Bangladesh	5.7	16.7
Sudan	5.6	20.7
Venezuela	5.1	21.3
Ethiopia	4.5	18.9
Spain	3.5	23.1
Australia	3	22.0
Japan	3	25.4
Canada	2.7	24.3
Denmark	2.2	25.6

b.

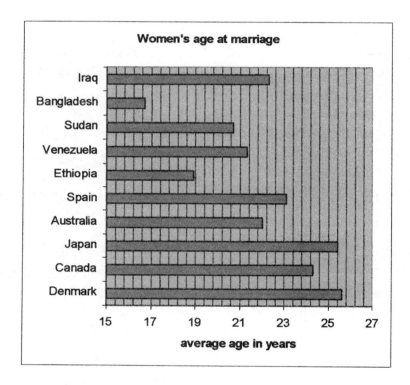

c.   Bangladesh; Denmark

d.   It seems that when women marry at a younger age, they have larger families. For example, Bangladesh has the youngest age at marriage and also has one of the largest family sizes. However, there must be other factors involved. For example, Iraq has the largest family size but does not have the youngest marriage age.

11. Blue area: 4 square in.
Red area: 4 square in.
Yellow area: 8 square in.

13. Red area: 8 square in.
Gray area: 12 square in.
Yellow area: 4 square in.
Light blue area: 4 square in.
Dark blue area: 8 square in.

15. $\dfrac{3}{8} + \dfrac{4}{5} = 0.375 + 0.8 = 1.175$

17. $\dfrac{11}{12} - \dfrac{1}{6} = 0.91\overline{6} - 0.1\overline{6} = 0.75$

19. $8 - 2\dfrac{1}{3} = 8 - 2.\overline{3} = 5.\overline{6}$

21. $4\dfrac{1}{6} - 2\dfrac{2}{3} = 4.1\overline{6} - 2.\overline{6} = 1.5$

23. $6\dfrac{5}{8} + 3\dfrac{2}{5} = 6.625 + 3.4 = 10.025$

Homework 2.5 B

1. a. Amount before the fair: A
    b. Amount Lupe has left: $A - 24$
    $\qquad\qquad\qquad\qquad\quad$ \$39
    $$A - 24 = 39$$
    c.
    $$A - 24 = 39$$
    $$\underline{+24 \quad +24}$$
    $$A = 63$$
    Lupe had \$63 before the Craft Fair.

3. a. Brenda's weight: $\qquad B$
    b. Danny's weight: $\qquad B + 32$
    $\qquad\qquad\qquad\qquad\qquad\quad$ 157
    $$B + 32 = 157$$
    c.
    $$B + 32 = 157$$
    $$\underline{-32 \quad -32}$$
    $$B = 125$$
    Brenda weighs 125 pounds.

5. a. Miranda's hourly wage: w
    b. Miranda's total earnings: $20w$
    $\qquad\qquad\qquad\qquad\qquad$ \$136
    $$20w = 136$$
    c.
    $$20w = 136$$
    $$\dfrac{20w}{20} = \dfrac{136}{20}$$
    $$w = 6.80$$
    Miranda's hourly wage is \$6.80.

7. a. Total profit: P
    b. Each member's share: $\dfrac{P}{8}$
    $\qquad\qquad\qquad\qquad\qquad$ 64
    $$\dfrac{P}{8} = 64$$
    c.
    $$\dfrac{P}{8} = 64$$
    $$8\left(\dfrac{P}{8}\right) = (64)8$$
    $$P = 512$$
    The total profit was \$512.

9.   Step 1: Number of hours Shant worked: w

Step 2: $w + 5 = 18$

Step 3:
$$w + 5 = 18$$
$$\underline{-5 \quad -5}$$
$$w = 13$$

Shant worked for 13 hours.

11.   Step 1: Rina's age: R

Step 2: $3R = 21$

Step 3:
$$3R = 21$$
$$\frac{3R}{3} = \frac{21}{3}$$
$$R = 7$$

Rina is 7 years old.

13.   Step 1: Number of disks Will has: n

Step 2: $n - 17 = 29$

Step 3:
$$n - 17 = 29$$
$$\underline{+17 \quad +17}$$
$$n = 46$$

Will has 46 disks.

15.   Step 1: Amount of Caroline's fortune: f

Step 2: $\dfrac{f}{4} = 350{,}000$

Step 3:
$$\frac{f}{4} = 350{,}000$$
$$4\left(\frac{f}{4}\right) = (350{,}000)4$$
$$f = 1{,}400{,}000$$

Caroline's fortune was $1,400,000.

17.   Step 1: Balance before check: B

Step 2: $B - 2378 = 1978$

Step 3:
$$B - 2378 = 1978$$
$$\underline{+2378 \quad +2378}$$
$$B = 4356$$

The balance was $4356.

19. a.

City	People per room	Deaths per 1000 Births
Singapore	1.2	7
Sydney	0.5	10
Hong Kong	1.6	7
Bangkok	3.2	27
Shanghai	2	14
Manila	3	36
Jakarta	3.4	45
Los Angeles	0.5	9
Seoul	2	12
Tokyo-Yokohama	0.9	5

b.   Jakarta; Jakarta

c.   Tokyo-Yokohama; Sydney and Los Angeles

d.   In general, crowded living conditions seem to be associated with higher infant mortality rates. Living conditions tend to be crowded and infant mortality rates tend to be high in cities affected by poverty and environmental challenges.

21. $\dfrac{7}{8}\left(\dfrac{4}{5}\right) = (0.875)(0.8) = 0.7$

23. $15\left(2\dfrac{1}{3}\right) = 15(2.\overline{3}) = 35$

25. $\dfrac{3}{8} \div \dfrac{5}{16} = 0.375 \div 0.3125 = 1.2$

27. $1 \div \dfrac{5}{8} = 1 \div 0.625 = 1.6$

29. $6 \div 2\dfrac{2}{5} = 6 \div 2.4 = 2.5$

Chapter 2 Review

1.   solution

3.   parallelogram

5.   formula

7.   area

9.   The value of a constant does not change, but a variable can take on different values.

11.   Terms are added or subtracted. Factors are multiplied together.

13.   Area is a measure of the amount of space inside a figure. Perimeter is the distance around the outside of a figure.

15.   Revenue is the amount of money earned in a business. Profit is the revenue minus the costs.

17.   The base of a triangle is the length of any one side. The height of a triangle is the distance from the base to the opposite vertex. The base of a parallelogram or trapezoid is the length of one of the parallel sides. The height is then the distance from the base to the opposite side.

19. $45 - t = 45 - 17 = 28$

21. $lw = (13)(6) = 78$

23. $\dfrac{560}{r} = \dfrac{560}{35} = 16$

25. $x + 3.7 = 4.4 + 3.7 = 8.1$

27. $4ac = 4(2)(3) = 24$

29.   Height of normal doorway: h; $5h$

31.   Price of book: p; $p - 8$

33.   Number of sheep: s; $\dfrac{156}{s}$

35.   Regular fare: r; $r - 250$

37.   Length of play: L; $L - 20$

39. $P = rW$; P = part, r = percentage rate, W = whole

41. $d = rt$; d = distance, r = rate, t = time

43. $e = \dfrac{m}{g}$; e = efficiency, m = miles, g = gallons

45. $I = Prt$; I = interest, P = principal, r = rate, t = time

47. $A = \dfrac{1}{2}bh$; A = area, b = base, h = height

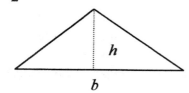

49. $y = 2x$

51. $j = i - 9$

53. $q = \dfrac{3}{2}p$

55. $v = \dfrac{60}{u}$

57.
$$19 + a = 32$$
$$\underline{-19 \qquad -19}$$
$$a = 13$$

Check:
$$19 + a = 32$$
$$19 + 13 = 32\,?$$
$$32 = 32$$

59.
$$22 = c - 11$$
$$\underline{+11 \qquad +11}$$
$$33 = c$$
$$c = 33$$

Check:
$$22 = c - 11$$
$$22 = 33 - 11\,?$$
$$22 = 22$$

61.
$$13h = 52$$
$$\dfrac{13h}{13} = \dfrac{52}{13}$$
$$h = 4$$

Check:
$$13h = 52$$
$$13(4) = 52\,?$$
$$52 = 52$$

63.
$$\dfrac{m}{4} = 8$$
$$4\left(\dfrac{m}{4}\right) = (8)4$$
$$m = 32$$

Check :
$$\dfrac{m}{4} = 8$$
$$\dfrac{32}{4} = 8\,?$$
$$8 = 8$$

65.
$$9.6 = 1.6p$$
$$\dfrac{9.6}{1.6} = \dfrac{1.6p}{1.6}$$
$$6 = p$$
$$p = 6$$

Check:
$$9.6 = 1.6p$$
$$9.6 = 1.6(6)\,?$$
$$9.6 = 9.6$$

67. $P = 5 + 8 + 5 + 8 = 26$ feet

69.
$$P = R - C$$
$$= 7 - 1.73$$
$$= \$5.27$$

71.
$$P = rW$$
$$= (0.05)(12{,}900)$$
$$= \$645$$

73.
$$A = lw$$
$$32 = l(4)$$
$$\dfrac{32}{4} = \dfrac{l(4)}{4}$$
$$8 = l$$
$$l = 8 \text{ meters}$$

75.
$$d = rt$$
$$121 = 55t$$
$$\dfrac{121}{55} = \dfrac{55t}{55}$$
$$2.2 = t$$
$$t = 2.2 \text{ hours}$$

77.
a. Age of older church: a
b. $173 = a - 238$
c.
$$173 = a - 238$$
$$\underline{+238 \qquad +238}$$
$$411 = a$$
The older church is 411 years old.

79.
a. Weight of son: w
b. $650 = w + 483$
c.
$$650 = w + 483$$
$$\underline{-483 \qquad -483}$$
$$167 = w$$
The son can weigh 167 pounds.

81.   a.   Size of jackpot: J

      b.

$$\frac{J}{6} = 2,535,000$$

      c.

$$\frac{J}{6} = 2,535,000$$

$$6\left(\frac{J}{6}\right) = (2,353,000)6$$

$$J = 15,210,000$$

The jackpot was \$15,210,000.

85.   a.   Number poiled: N

      b. $0.65N = 663$

      c.

$$0.65N = 663$$

$$\frac{0.65N}{N} = \frac{663}{0.65}$$

$$N = 1020$$

1020 voters were polled.

87.   Perimeter: $5 + 4 + 1 + 1 + 1 + 2 + 3 + 3 = 20$ cm

      Area: 18 square cm

89.

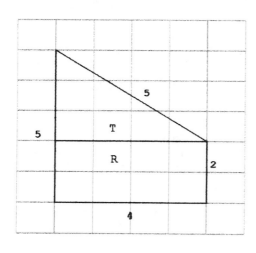

83.   a.   Time to swing back and forth: t

      b. $4 = 5t$

      c.

$$4 = 5t$$

$$\frac{4}{5} = \frac{5t}{5}$$

$$0.8 = t$$

The pendulum takes 0.8 seconds to swing back and forth.

Perimeter: $5 + 5 + 2 + 4 = 16$ cm

Area = (area of triangle) +   (area of rectangle)

$$= \left(\frac{1}{2}bh\right) + (lw)$$

$$= \left(\frac{1}{2} \cdot 4 \cdot 3\right) + (4 \cdot 2)$$

$$= 6 + 8$$

$$= 14 \text{ square cm}$$

91.

$$A = bh$$

$$= (2)(4)$$

$$= 8 \text{ square cm}$$

93.

Area = (area of large square) − (area of missing triangle)

$$= (lw) - \left(\frac{1}{2}bh\right)$$
$$= (5 \cdot 4) - \left(\frac{1}{2} \cdot 3 \cdot 3\right)$$
$$= 20 - 4.5$$
$$= 15.5 \text{ square cm}$$

95.   3.11

97.

$$\frac{5}{6} = 0.8333\cdots$$

$\frac{5}{6}$ is larger.

99.   4.79

101. $8\frac{1}{8}\% = 8.125\% = 0.08125$

103.   2.09

105.   1.00

107. $\frac{11}{13} = 11 \div 13 = 0.8461\cdots \approx 0.846$

109. $2\frac{1}{6} = 1 \div 6 + 2 = 2.1666\cdots \approx 2.167$

111. $2\frac{3}{4} \times 3\frac{1}{3} = 2.75 \times 3.\overline{3} = 9.1\overline{6}$

113. $20 \div 2\frac{2}{5} = 20 \div 2.4 = 8.\overline{3}$

Homework 3.1 A

1. −$100

3. 2 inches

5. −450 feet

7. −$3,000,000

9. −384

11. −3°

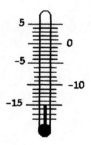

13. 0°

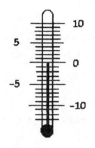

15. −15°

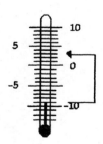

17. 2°

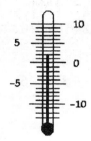

19.

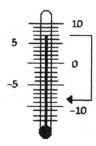

Start at −9°. Then count up 12°. The temperature at noon was 3°.

21.

Start at 7°. Then count down 15°. The temperature at 4 p.m. was −8°.

23.

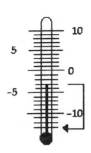

Start at −3°. Then count down 10°.
The temperature at noon was −13°.

25. a. −3° C
     b. January has the coldest low
         temperature of −11°. July has
         the warmest low temperature of
         17°.
     c. 13 − (−9) = 22° C

27.

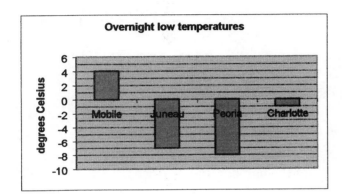

29. perimeter

31. perimeter

33. area

35. area

37. perimeter

39.
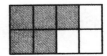

41.

43.

45.

47.

Homework 3.1 B

1.

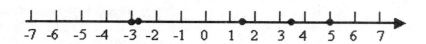

3.

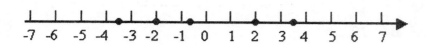

5. $6.5 > 6.07$ 7. $-12 < -2$

9. $-\dfrac{3}{4} < -\dfrac{1}{2}$

11.   True; Since -3 is to the right of -5 on the number line, $-3 > -5$.

13.   False; Since -7 is to the left of 6 on the number line, $-7 < 6$.

15.   True; Since -12 is to the left of -9 on the number line, $-12 < -9$.

17.   False; Since $-\dfrac{1}{2}$ is to the right of $-1\dfrac{1}{2}$ on the number line, $-\dfrac{1}{2} > -1\dfrac{1}{2}$.

19.   False; Since 0.1 is to the right of 0.01 on the number line, $0.1 > 0.001$.

21.   2.0001, 3, 4, 15   23. $-3.1, -4, -1000, -12$

25. $-4.0001, -5, -6, -100$ 27. $-(4) = -4$

29. $-(-15) = 15$ 31. $-(-2) = 2$

33. $-[-(-7)] = -[7] = -7$ 35.   Aunt Ethel's age: E
                                                    $E > 65$

37.   Temperature of liquid copper: T 39.   Score on physics test: S
        $T > 1083$ $S < 85$

41.   Depth: d
        $d > -50$

43.

Country	Railroads in 1900 (Thousands of Miles)	Railroads in 2000 (Thousands of Miles)	Change in Miles
United States	193	149	$149 - 193 = -44$
Russia	33	54	$54 - 33 = 21$
India	26	38	$38 - 26 = 12$
France	24	21	$21 - 24 = -3$
Britain	19	11	$11 - 19 = -8$

45.

47.

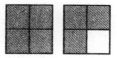

49.

51.

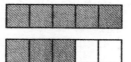

53.

Homework 3.2 A

1. $(+4) + (+5) = 9$

3. $(-2) + (-6) = -8$

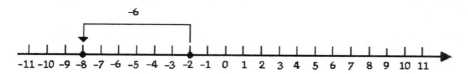

5. $(-3) + (+7) = 4$

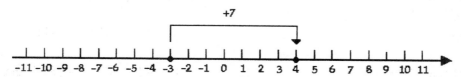

7. $(+10) + (-12) = -2$

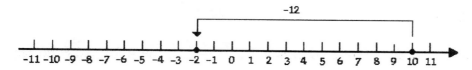

9. $(-8) + (+4) = -4$

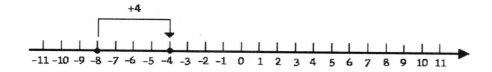

11.   If you start at a negative number on the number line and go to the left, you will be at another negative number.

13.   If the balance in your checking account was −$100, and then you made a deposit of $200, your balance is $-100 + 200 = \$100$.

15. $(+7) + (+9) = +(7+9) = 16$ 17. $(-5) + (-6) = -(5+6) = -11$

19. $(-12) + (-17) = -(12+17) = -29$ 21. $(-8) + (-8) = -(8+8) = -16$

23. $-25 + (-50) = -(25 + 50) = -75$

25. $(+12) + (-5) = +(12 - 5) = 7$

27. $20 + (-7) = +(20 - 7) = 13$

29. $(-16) + (+24) = +(24 - 16) = 8$

31. $(-6) + (+13) = +(13 - 6) = 7$

33. $(-5) + (+5) = 5 - 5 = 0$

35. $-8 + (+2) = -(8 - 2) = -6$

37. $-17 + (+4) = -(17 - 4) = -13$

39. $(+9) + (-18) = -(18 - 9) = -9$

41. $21 + (-15) = +(21 - 15) = 6$

43. $-16 + (-18) = -(16 + 18) = -34$

45. $-29 + 7 = -(29 - 7) = -22$

47. $(-38) + (-76) = -114$

49. $(-0.5) + (-0.3) = -0.8$

51. $-136 + (-245) = -381$

53. $13.8 + (-18.3) = -4.5$

55. $-2.05 + 6.54 = 4.49$

57. $(-27) + (+6) = -21$ pounds

59. $(+1200) + (-1400) = -200$ feet

61. $(-5000) + (-1800) = -6800$ feet

63. $(-9) + (+9) = 9 - 9 = 0$

65. $(+63) + (-63) = 63 - 63 = 0$

67. $(-0.2) + (0.2) = 0.2 - 0.2 = 0$

69. If you deposit \$100 into your checking account and then write a check for \$100, the net change in your balance is 0.

71. $1 - \dfrac{1}{3} = \dfrac{3}{3} - \dfrac{1}{3} = \dfrac{2}{3}$

73. $1 - \dfrac{2}{5} = \dfrac{5}{5} - \dfrac{2}{5} = \dfrac{3}{5}$

75. $1 - \dfrac{5}{9} = \dfrac{9}{9} - \dfrac{5}{9} = \dfrac{4}{9}$

77. $1 - \dfrac{7}{12} = \dfrac{12}{12} - \dfrac{7}{12} = \dfrac{5}{12}$

79. $1 - \dfrac{3}{4} = \dfrac{4}{4} - \dfrac{3}{4} = \dfrac{1}{4}$

Homework 3.2 B

1. $(-700) + 250 = -450$ feet

3. $(-5600) + (-3450) = -\$9050$

5. $1876.47 + (-1595.62) = \$280.85$

7. $8 + (-6) = +(8 - 6) = 2$

9. $-8 + 6 = -(8 - 6) = -2$

11. $-20 + (-30) = -(20 + 30) = -50$

13.   a. $1.3 million, −$0.8 million, −$0.2 million, $2.1 million

        b. $(1.3) + (−0.8) + (−0.2) + (2.1) = 2.4 million

15.   a.

Month	Jan	Feb	Mar	Apr	May	Jun	Jul	Aug	Sep	Oct	Nov	Dec
Relative water level (feet)	3	6	8	7	6	4	1	−3	−5	−7	−6	−4
Water depth (feet)	37	40	42	41	40	38	35	31	29	27	28	30

        b.

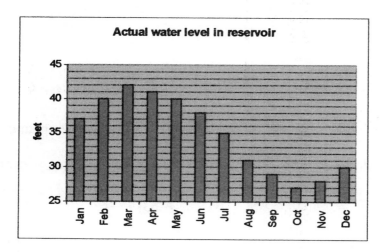

17. $m + p = (−8) + (−4) = −12$

19. $t + m = 6 + (−8) = −2$

21.

$$m + p + t = (−8) + (−4) + 6$$
$$= (−12) + 6$$
$$= −6$$

23.

$$t + p + t = 6 + (−4) + 6$$
$$= 2 + 6$$
$$= 8$$

25.

$$x − 9 = −2$$
$$\underline{+9 \quad +9}$$
$$x = 7$$

27.

$$a − 4 = −18$$
$$\underline{+4 \quad +4}$$
$$a = −14$$

29.

$$−5 + v = −14$$
$$\underline{+5 \qquad +5}$$
$$v = −9$$

31.

$$−12 + g = −12$$
$$\underline{+12 \qquad +12}$$
$$g = 0$$

33.

$$−5.7 + n = −8.4$$
$$\underline{+5.7 \qquad +5.7}$$
$$n = −2.7$$

35.

$$s − 83.2 = −12.6$$
$$\underline{+83.2 \quad +83.2}$$
$$s = 70.6$$

37.

$$j − 0.04 = −0.6$$
$$\underline{+0.04 \quad +0.04}$$
$$j = −0.56$$

Lesson 3.2

39. $x + y = 2.5 + (-7.2) = -4.7$

41. $z + x = (-4.8) + 2.5 = -2.3$

43.
$$\begin{aligned} z + z + z &= (-4.8) + (-4.8) + (-4.8) \\ &= (-9.6) + (-4.8) \\ &= -14.4 \end{aligned}$$

45.
$$\begin{aligned} x + x + y &= 2.5 + 2.5 + (-7.2) \\ &= 5 + (-7.2) \\ &= -2.2 \end{aligned}$$

47. $6p$

49. $\dfrac{1200}{v}$

51. $1800 - c$

53. $m - 28$

55. $\dfrac{5000}{w}$

57. $\dfrac{9}{5} = \dfrac{5+4}{5} = \dfrac{5}{5} + \dfrac{4}{5} = 1\dfrac{4}{5}$

59. $\dfrac{11}{6} = \dfrac{6+5}{6} = \dfrac{6}{6} + \dfrac{5}{6} = 1\dfrac{5}{6}$

61. $\dfrac{10}{7} = \dfrac{7+3}{7} = \dfrac{7}{7} + \dfrac{3}{7} = 1\dfrac{3}{7}$

63. $\dfrac{5}{4} = \dfrac{4+1}{4} = \dfrac{4}{4} + \dfrac{1}{4} = 1\dfrac{1}{4}$

65. $\dfrac{13}{8} = \dfrac{8+5}{8} = \dfrac{8}{8} + \dfrac{5}{8} = 1\dfrac{5}{8}$

Homework 3.3 A

1. $8 - (-4) = 8 + (+4) = 12$

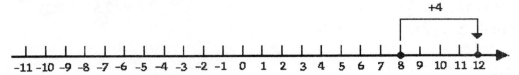

3. $-3 - (+8) = -3 + (-8) = -11$

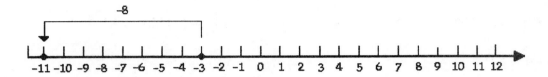

5. $11 - (+5) = 11 + (-5) = 6$

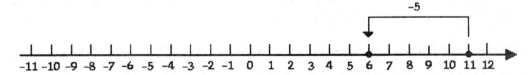

7. $-9 - (-2) = -9 + (+2) = -7$

9. Taking away a debt has the effect of adding to net worth.

11. $12 - (+3) = 12 - 3 = 9$

13. $5 - (+11) = 5 + (-11) = -(11 - 5) = -6$

15. $-6 - (+9) = -6 + (-9) = -(6 + 9) = -15$

17. $-8 - (+4) = -8 + (-4) = -(8 + 4) = -12$

19. $-5 - (+5) = -5 + (-5) = -(5 + 5) = -10$

21. $7 - (+7) = 7 - 7 = 0$

23. $9 - (-8) = 9 + (+8) = 17$

25. $17 - (-20) = 17 + (+20) = 37$

27. $-13 - (+9) = -13 + (-9) = -(13 + 9) = -22$

29. $-12 - (+19) = -12 + (-19) = -(12 + 19) = -31$

31. $-17 - (+17) = -17 + (-17) = -(17 + 17) = -34$

33. $-16-(-5) = -16+(+5) = -(16-5) = -11$

35. $-22-(-28) = -22+(+28) = 28-22 = 6$

37. $-30-(-30) = -30+(+30) = 30-30 = 0$

39. a. Round 1: $4-7 = -3$
    Round 2: $2-4 = -2$
    Round 3: $6-3 = 3$
    b. Final Score $= (-3)+(-2)+3 = -5+3 = -2$

41. $23.8-(-82.9) = 106.7$ 43. $-249-(-152) = -97$

45. $0.012-(-0.39) = 0.402$ 47. $-1.05-(-0.15) = -0.9$

49. a. "Sixteen minus positive four"; $16-(+4) = 16-4 = 12$
    "Sixteen plus negative four"; $16+(-4) = 16-4 = 12$
    b. The expression can be interpreted either way. Both interpretations give the same value of 12. Preferences will vary.

51. a. "Negative three minus positive nine"; $-3-(+9) = -3+(-9) = -(9+3) = -12$
    "Negative three plus negative nine"; $-3+(-9) = -(3+9) = -12$
    b. The expression can be interpreted either way. Both interpretations give the same value of -12. Preferences will vary.

53. a. $8-15 = 8+(-15)$
    $8-15 = 8-(+15)$
    b. Answers will vary.

55. a. $-5-7 = -5+(-7) = -(5+7) = -12$
    b. $2-9 = 2+(-9) = -(9-2) = -7$

57. $13+(-5) = 13-5 = 8$ 59. $-27+(-13) = -(27+13) = -40$

61. $35-11 = 24$

63. $-40-30 = -40+(-30) = -(40+30) = -70$

65. $-18+53 = 53-18 = 35$ 67. $-3-(-5) = -3+(+5) = 5-3 = 2$

69. $3-5 = 3+(-5) = -(5-3) = -2$

71. Eight; $2 = \dfrac{8}{4}$

73.　Nine; $3 = \dfrac{9}{3}$

75.　Eighteen; $3 = \dfrac{18}{6}$

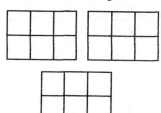

77.　Ten; $5 = \dfrac{10}{2}$

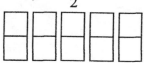

79.　Twenty-four; $3 = \dfrac{24}{8}$

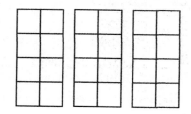

Homework 3.3 B

1.　$8 - 10 - 6 = -2 - 6 = -8$

3.
$$-3 - (-18) - 9 = -3 + 18 - 9$$
$$= 15 - 9$$
$$= 6$$

5.　$6 - (-6) - 6 = 6 + 6 - 6 = 12 - 6 = 6$

7.
$$-2 + 3 - (-4) - 5 = -2 + 3 + 4 - 5$$
$$= 1 + 4 - 5$$
$$= 5 - 5$$
$$= 0$$

9.　$-12 - 12 - 12 = -24 - 12 = -36$

11.　$-a = -(-3) = 3$

13.　$-(-c) = -(-(-7)) = -(7) = -7$

15.　$-b - a = -(8) - (-3) = -8 + 3 = -5$

17.
$$b - c - a = (8) - (-7) - (-3)$$
$$= 8 + 7 + 3$$
$$= 15 + 3$$
$$= 18$$

19.
$$a - b + c = (-3) - (8) + (-7)$$
$$= -3 - 8 - 7$$
$$= -11 - 7$$
$$= -18$$

21. a.

Year	1995	1996	1997	1998	1999
Change in exports (Billions of dollars)	$4.6 - 4.0$ $= 0.6$	$4.7 - 4.6$ $= 0.1$	$6.4 - 4.7$ $= 1.9$	$6.3 - 6.4$ $= -0.1$	$5.4 - 6.3$ $= -1.1$

b.

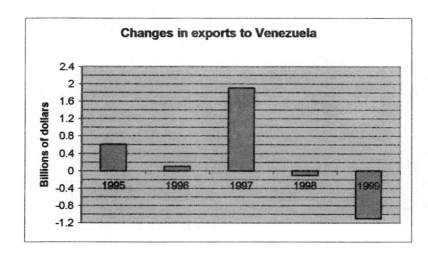

23. a.

Year	1996	1997	1998	1999	2000
Aircraft	$468 + 286$ $= 754$	$754 - 186$ $= 568$	$568 + 88$ $= 656$	$656 - 265$ $= 391$	$391 + 194$ $= 585$

b. 1996

25.
$$13 + a = 9$$
$$\underline{-13 \qquad -13}$$
$$a = -4$$

Check: $13 + a = 9$
$13 + (-4) = 9$
$9 = 9$

27.
$$p + 12 = -6 - 4$$
$$p + 12 = -10$$
$$\underline{-12 \quad -12}$$
$$p = -22$$

Check:
$p + 12 = -6 - 4$
$(-22) + 12 = -6 - 4$
$-10 = -10$

29
$$h + 16 = 12 - (-7)$$
$$h + 16 = 12 + 7$$
$$h + 16 = 19$$
$$\underline{-16 \quad -16}$$
$$h = 3$$

Check:
$h + 16 = 12 - (-7)$
$3 + 16 = 12 + 7$
$19 = 19$

31.
$$m - 6 = -11$$
$$\underline{+6 \quad +6}$$
$$m = -5$$

Check: $m - 6 = -11$
$(-5) - 6 = -11$
$-11 = -11$

33.
$$-36 + t = -22$$
$$\underline{+36 \qquad +36}$$
$$t = 14$$

Check: $-36 + t = -22$
$-36 + 14 = -22$
$-22 = -22$

35.
$$41 + u = -33 + 14$$
$$41 + u = -19$$
$$\underline{-41 \qquad -41}$$
$$u = -60$$

Check:
$41 + u = -33 + 14$
$41 + (-60) = -19$
$-19 = -19$

37.
$$c - 6.8 = -3.9 \qquad \text{Check: } c - 6.8 = -3.9$$
$$\underline{+6.8 \quad +6.8} \qquad\qquad 2.9 - 6.8 = -3.9$$
$$c = 2.9 \qquad\qquad\qquad -3.9 = -3.9$$

39.
$$f - (-4.8) = -2.3 \qquad \text{Check:}$$
$$f + \quad 4.8 = -2.3 \qquad\qquad f - (-4.8) = -2.3$$
$$\underline{-4.8 \quad -4.8} \qquad (-7.1) - (-4.8) = -2.3$$
$$f = -7.1 \qquad\qquad -7.1 + 4.8 = -2.3$$
$$-2.3 = -2.3$$

41.
$$v - t = (-1.4) - (2.1)$$
$$= (-1.4) + (-2.1)$$
$$= -3.5$$

43.
$$v - t + w = (-1.4) - (2.1) + (-0.2)$$
$$= (-1.4) + (-2.1) + (-0.2)$$
$$= (-3.5) + (-0.2)$$
$$= -3.7$$

45.
$$-w + t + v = -(-0.2) + (2.1) + (-1.4)$$
$$= (0.2) + (2.1) + (-1.4)$$
$$= 2.3 + (-1.4)$$
$$= 0.9$$

47.

x	$x + 4$
-7	$(-7) + 4 = \mathbf{-3}$
-5	$(-5) + 4 = \mathbf{-1}$
-3	$(-3) + 4 = \mathbf{1}$
-1	$(-1) + 4 = \mathbf{3}$

49.

h	$h - 5$
-8	$(-8) - 5 = \mathbf{-13}$
-5	$(-5) - 5 = \mathbf{-10}$
-2	$(-2) - 5 = \mathbf{-7}$
2	$(2) - 5 = \mathbf{-3}$

51. Since there are 3 feet in one yard, in k yards there are $3k$ feet.

$$1 \text{ yard } = 3 \text{ feet}$$
$$k \text{ yards } = 3k \text{ feet}$$

53. Since there are 2000 pounds in one ton, in t tons there are $2000t$ pounds.

$$1 \text{ ton } = 2000 \text{ pounds}$$
$$t \text{ tons } = 2000t \text{ pounds}$$

55. Since there are 4 quarts in one gallon, in g gallons there are $4g$ quarts.

$$1 \text{ gallon } = 4 \text{ quarts}$$
$$g \text{ gallons } = 4g \text{ quarts}$$

57. Since there are 16 ounces in one pound, in z ounces there are $\frac{z}{16}$ pounds.

$$16 \text{ ounces } = 1 \text{ pound}$$
$$1 \text{ ounce } = \frac{1}{16} \text{ pound}$$
$$z \text{ ounces } = \frac{z}{16} \text{ pounds}$$

59. Since there are 12 inches in one foot, in n inches there are $\frac{n}{12}$ feet.

$$12 \text{ inches } = 1 \text{ foot}$$
$$1 \text{ inch } = \frac{1}{12} \text{ foot}$$
$$n \text{ inches } = \frac{n}{12} \text{ feet}$$

61. $\dfrac{8}{4} = 2$;

There are 2 groups of four fourths.

63. $\dfrac{9}{3} = 3$;

There are 3 groups of three thirds.

65. $\dfrac{8}{2} = 4$;

There are 4 groups of two halves.

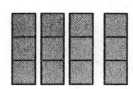

67. $\dfrac{5}{5} = 1$;

There is one group of five fifths.

69. $\dfrac{12}{3} = 4$;

There are 4 groups of three thirds.

These problems can be worked by dividing the numerator by the denominator. For example, $\dfrac{12}{6} = 12 \div 6 = 2$.

Homework 3.4 A

1. $3(-6) = -18$ 3. $-9(-2) = 18$ 5. $-7 \cdot 4 = -28$

7. $-36 \div 4 = -9$ 9. $-56 \div (-8) = 7$ 11. $24 \div (-8) = -3$

13. $0 \div (-12) = 0$ 15. $-12 \div 0$ undefined   17. $12 \div (-12) = -1$

19. $\dfrac{-84}{-6} = 14$ 21. $\dfrac{96}{-6} = -16$ 23. $\dfrac{0}{-7} = 0$

25. When we interpret multiplication as repeated addition, the positive number tells us how many times to add the negative number. For example, $3(-2) = (-2) + (-2) + (-2) = -6$. Since the sum of negative numbers is negative, the product of a positive and a negative number is also negative.

27. a. $-6(-8) = 48$
    b. $-6 - 8 = -14$

29. a. $12 - (-4) = 12 + 4 = 16$
    b. $12 \div (-4) = -3$

31. a. $-40 - (-8) = -40 + 8 = -32$
    b. $-40 \div (-8) = 5$

33. a. $-3 + (-8) = -11$
    b. $-3(-8) = 24$

35. a. $-9(7) = -63$
    b. $-9 + 7 = -2$

37. $-40 \div 8 = -5 \Rightarrow 8(-5) = -40$

39. $\dfrac{56}{-4} = -14 \Rightarrow -4(-14) = 56$

41. $\dfrac{a}{-7} = 28 \Rightarrow -7(28) = a$

43. $-18(-23) = 414$

45. $-434 \div (-14) = 31$

47. $\dfrac{288}{-16} = -18$

49. $(-0.3)(26.1) = -7.83$

51.

$$e = \frac{m}{g}$$

$$26 = \frac{380}{g}$$

$$g(26) = \left(\frac{380}{g}\right)g$$

$$26g = 380$$

$$\underline{a}$$

53.   <u>b</u>   55.   <u>a</u>

57. $1\frac{3}{4} = $ 1 whole $+\frac{3}{4} = \frac{7}{4}$ 59. $3\frac{1}{3} = $ 3 wholes $+\frac{1}{3} = \frac{10}{3}$

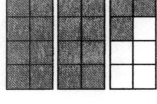   

61. $2\frac{3}{8} = $ 2 wholes $+\frac{3}{8} = \frac{19}{8}$ 63. $4\frac{1}{2} = $ 4 wholes $+\frac{1}{2} = \frac{9}{2}$

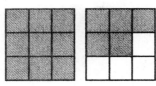   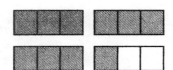

65. $1\frac{5}{9} = $ 1 whole $+\frac{5}{9} = \frac{14}{9}$

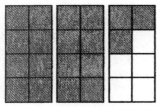

The denominator of the improper fraction is the same as the denominator in the mixed number. To find the numerator of the improper fraction, multiply the whole number times the denominator and add the numerator. For example, $2\frac{5}{6} = \frac{2\cdot6+5}{6} = \frac{17}{6}$.

Homework 3.4 B

1. $(-3)(-4)(-6) = (12)(-6) = -72$

3.
$$(-5)(-5)(-5)(-5) = (25)(-5)(-5)$$
$$= (-125)(-5)$$
$$= 625$$

5. $-80 \div 8 \div (-2) = -10 \div (-2) = 5$

7.
$$12(-20) \div (-5)(-2) = -240 \div (-5)(-2)$$
$$= 48(-2)$$
$$= -96$$

9.
$$-10 \div 10(-10) \div (-10) = -1(-10) \div (-10)$$
$$= 10 \div (-10)$$
$$= -1$$

11. $\quad -3m = -3(8) = -24$

13. $\quad \dfrac{26}{h} = \dfrac{26}{-2} = -13$

15. $\quad \dfrac{10h}{m} = \dfrac{10(-2)}{8} = \dfrac{-20}{8} = -\dfrac{5}{2}$

17.
$$d \cdot d \cdot d = (-4)(-4)(-4)$$
$$= (16)(-4)$$
$$= -64$$

19.
$$d + d + d = (-4) + (-4) + (-4)$$
$$= (-8) + (-4)$$
$$= -12$$

21. $\quad \dfrac{q}{8} = \dfrac{-0.8}{8} = -0.1$

23.
$$k \div q \div d = (-2.4) \div (-0.8) \div (-4)$$
$$= (3) \div (-4)$$
$$= -0.75$$

25.
$$2qk = 2(-0.8)(-2.4)$$
$$= (-1.6)(-2.4)$$
$$= 3.84$$

27. $\quad \dfrac{0}{q} = \dfrac{0}{-0.8} = 0$

29. a. $\quad -\$1$
    b. $\quad 50(-1) = -\$50$
    c. $\quad$ AmExp: $50(0.25) = \$12.50$
        Disney: $50(-0.25) = -\$12.50$
        ExxonMob: $50(-1.5) = -\$75$
        GnMotr: $50(-0.75) = -\$37.50$
    d. $\quad (-50) + (12.50) + (-12.50) + (-75) + (-37.50) = -\$162.50$

31.
$$8q = -24$$
$$\dfrac{8q}{8} = \dfrac{-24}{8}$$
$$q = -3$$

33.
$$-2r = -18$$
$$\dfrac{-2r}{-2} = \dfrac{-18}{-2}$$
$$r = 9$$

35.
$$-6y = 42$$
$$\dfrac{-6y}{-6} = \dfrac{42}{-6}$$
$$y = -7$$

37.
$$\dfrac{p}{9} = -9$$
$$9\left(\dfrac{p}{9}\right) = (-9)9$$
$$p = -81$$

39.

$$\frac{s}{-5} = -15$$

$$(-5)\left(\frac{s}{-5}\right) = (-15)(-5)$$

$$s = 75$$

41.

$$\frac{f}{-4} = 18$$

$$(-4)\left(\frac{f}{-4}\right) = (18)(-4)$$

$$f = -72$$

43.

$$\frac{h}{10} = -2.8$$

$$10\left(\frac{h}{10)}\right) = (-2.8)(10)$$

$$h = -28$$

45.

$$\frac{z}{-0.001} = 59$$

$$(-0.001)\left(\frac{z}{-0.001}\right) = 59(-0.001)$$

$$z = -0.059$$

47.

$$1.5c = -90$$

$$\frac{1.5c}{1.5} = \frac{-90}{1.5}$$

$$c = -60$$

49.

$$-0.3b = -0.12$$

$$\frac{-0.3b}{-0.3} = \frac{-0.12}{-0.3}$$

$$b = 0.4$$

51.

$$x + 8 = -3$$

$$\frac{-8 \quad -8}{x = -11}$$

53.

$$-3x = -24$$

$$\frac{-3x}{-3} = \frac{-24}{-3}$$

$$x = 8$$

55.

$$x - 12 = -2$$

$$\frac{+12 \quad +12}{x = 10}$$

57.

$$\frac{x}{-2} = -8$$

$$(-2)\left(\frac{x}{-2}\right) = (-8)(-2)$$

$$x = 16$$

59. Step 1: Let m be the number of male chemistry majors.

Step 2: $\dfrac{5}{2} = \dfrac{m}{18}$

Step 3:

$$\frac{5}{2} = \frac{m}{18}$$

$$2m = (5)(18)$$

$$2m = 90$$

$$\frac{2m}{2} = \frac{90}{2}$$

$$m = 45$$

There are 45 male chemistry majors.

61. Step 1: Let L be the distance to the lodge.

Step 2: $4.1 = L + 1.4$

Step 3:
$$4.1 = L + 1.4$$
$$\underline{-1.4 \qquad -1.4}$$
$$2.7 = L$$

It is 2.7 miles to the lodge.

63. Step 1: Let n be the number of voters surveyed.

Step 2: $304 = (0.38)n$

Step 3:
$$304 = (0.38)n$$
$$\frac{304}{0.38} = \frac{(0.38)n}{0.38}$$
$$800 = n$$

800 voters were surveyed.

65.

b	$-4b$
-6	$-4(-6) = 24$
-4	$-4(-4) = 16$
4	$-4(4) = -16$
6	$-4(6) = -24$

67.

n	$\dfrac{-12}{n}$
-18	$\dfrac{-12}{-18} = \dfrac{2}{3}$
-8	$\dfrac{-12}{-8} = \dfrac{3}{2}$
8	$\dfrac{-12}{8} = -\dfrac{3}{2}$
18	$\dfrac{-12}{18} = -\dfrac{2}{3}$

69. $\dfrac{5}{3} = 1 \text{ whole} + \dfrac{2}{3} = 1\dfrac{2}{3}$

71. $\dfrac{9}{4} = 2 \text{ wholes} + \dfrac{1}{4} = 2\dfrac{1}{4}$

73. $\dfrac{19}{8} = 2 \text{ wholes} + \dfrac{3}{8} = 2\dfrac{3}{8}$

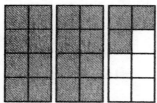

75. $\dfrac{16}{3} = 5 \text{ wholes} + \dfrac{1}{3} = 5\dfrac{1}{3}$

77. $\dfrac{10}{4} = 2 \text{ wholes} + \dfrac{2}{4} = 2\dfrac{2}{4} = 2\dfrac{1}{2}$

Divide the numerator by the denominator. Use the quotient as the whole number part of the mixed number and the remainder as the numerator of the fraction part of the mixed number. The denominator will be the same as the denominator of the fraction. Then simplify the fraction if possible. For example, $\dfrac{15}{6} = 15 \div 6 = 2 \text{ R } 3 = 2\dfrac{3}{6} = 2\dfrac{1}{2}$.

Homework 3.5

1. $n + 12 = 5$

3. $-12 + n = -5$

5. $n - 3 = -30$

7. $\dfrac{n}{3} = -30$

9. a. Temperature last night: T
    b. Start with the temperature last night and add the change in temperature to get the temperature this afternoon.

    $$\begin{pmatrix}\text{temperature}\\ \text{last night}\end{pmatrix} + \begin{pmatrix}\text{change in}\\ \text{temperature}\end{pmatrix} = \begin{pmatrix}\text{temperature}\\ \text{this afternoon}\end{pmatrix}$$

    $$T + 17 = -6$$

    c.

    $$T + 17 = -6$$
    $$\underline{-17 \quad -17}$$
    $$T = -23$$

    The temperature last night was $-23°$.

11. a. Balance yesterday: B
     b. Start with the balance yesterday and add her deposit to get the balance today.

     $$\begin{pmatrix}\text{balance}\\ \text{yesterday}\end{pmatrix} + (\text{deposit}) = \begin{pmatrix}\text{balance}\\ \text{today}\end{pmatrix}$$

     $$B + 132 = 74$$

     c.

     $$B + 132 = 74$$
     $$\underline{-132 \quad -132}$$
     $$B = -58$$

     Her balance yesterday was $-\$58$.

13. a. Elevation before lunch: E
     b. Start with his elevation before lunch and add his climb to get his elevation after lunch.

     $$\begin{pmatrix}\text{elevation}\\ \text{before lunch}\end{pmatrix} + \begin{pmatrix}\text{elevation}\\ \text{climbed}\end{pmatrix} = \begin{pmatrix}\text{elevation}\\ \text{after lunch}\end{pmatrix}$$

     $$E + 248 = 109$$

     c.

     $$E + 248 = 109$$
     $$\underline{-248 \quad -248}$$
     $$E = -139$$

     His elevation was -139 feet before lunch.

15. a.   Year Socrates was born: y
     b.   Start with his birth year and add his age to get the year he died.

$$\begin{pmatrix} \text{birth} \\ \text{year} \end{pmatrix} + (\text{age}) = \begin{pmatrix} \text{death} \\ \text{year} \end{pmatrix}$$
$$y + 71 = -399$$

     c.

$$y + 71 = -399$$
$$\underline{-71 \quad -71}$$
$$y = -470$$

Socrates was born in 470 BC.

17. a.   Yards gained: y
     b.   Start with Thad's yardage before the game and add his yardage during the game to get his yardage after the game.

$$\begin{pmatrix} \text{yardage} \\ \text{before} \\ \text{game} \end{pmatrix} + \begin{pmatrix} \text{yardage} \\ \text{during} \\ \text{game} \end{pmatrix} = \begin{pmatrix} \text{yardage} \\ \text{after} \\ \text{game} \end{pmatrix}$$
$$-18 + y = -26$$

     c.

$$-18 + y = -26$$
$$\underline{+18 \qquad +18}$$
$$y = -8$$

Thad gained -8 yards. In other words, he lost 8 yards.

19. a.   Change in elevation: c
     b.   Start with the initial elevation and add the change in elevation to get the final elevation.

$$\begin{pmatrix} \text{initial} \\ \text{elevation} \end{pmatrix} + \begin{pmatrix} \text{change} \\ \text{in elevation} \end{pmatrix} = \begin{pmatrix} \text{final} \\ \text{elevation} \end{pmatrix}$$
$$-286 + c = -159$$

     c.

$$-286 + c = -159$$
$$\underline{+286 \qquad +286}$$
$$c = 127$$

The change in the Yellow Submarine's elevation was 127 feet.

21. $-18 - 12 = -18 + (-12) = -30$

23. $\dfrac{-72}{6} = -12$

25. $-21 - (-17) = -21 + 17 = -4$

27. $-16(-4) = 64$

29. $-15 \div 0 = \dfrac{-15}{0}$ undefined

31.   a.   −$1.50; Each share lost $1.50 in value.

      b.   Starting price: s

$$\left(\begin{array}{c}\text{starting}\\\text{price}\end{array}\right)+\left(\begin{array}{c}\text{change}\\\text{in price}\end{array}\right)=\left(\begin{array}{c}\text{final}\\\text{price}\end{array}\right)$$

$$s-1.50=81$$
$$\underline{+1.50 \quad +1.50}$$
$$s=82.50$$

The starting price was $82.50.

33.

Divide the dots into thirds and circle two of the three thirds. Then count the dots enclosed.

$\frac{2}{3}$ of a day $= 16$ hours

35.

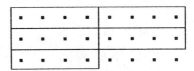

Divide the dots into sixths and circle five of the six sixths. Then count the dots enclosed.

$\frac{5}{6}$ of a day $= 20$ hours

37.

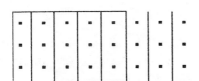

Divide the dots into eighths and circle five of the eight eighths. Then count the dots enclosed.

$\frac{5}{8}$ of a day $= 15$ hours

39.

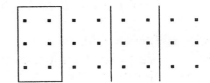

Divide the dots into fourths and circle one of the four fourths. Then count the dots enclosed.

$\frac{1}{4}$ of a day $= 6$ hours

41.

Divide the dots into eighths and circle one of the eight eighths. Then count the dots enclosed.

$\frac{1}{8}$ of a day $= 3$ hours

Chapter 3 Review

1. Temperature, altitude, profit

3. In an equation, the two expressions are separated by an equal sign. In an inequality, the two expressions are separated by a less than or greater than sign.

5. Answers may vary.

7. Answers may vary.

9. Undefined

11. $>$

13. $<$

15. $=$

17. a. Zoe's rent: R
    b. $R < 650$

19. a. Temperature last night: T
    b. $T < -15$

21. $-2 + (-4) = -6$

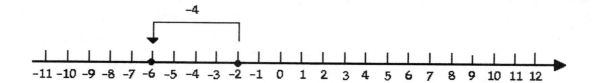

23. $-5 - (-7) = -5 + 7$

25. Negative

27. Zero

29. $-8(-4) = 32$

31. $\dfrac{-36}{4} = -9$

33. $-11 - 23 = -11 + (-23) = -34$

35. $-15 - (-8) = -15 + 8 = -7$

37. $18 - 20 = -2$

39. $-12 + 12 = 12 - 12 = 0$

41. $-3 + (-8) - 14 = -11 - 14 = -25$

43.
$$-20 - 5 + 7 = -25 + 7$$
$$= -(25 - 7)$$
$$= -18$$

45. $-3(-9)(-2) = 27(-2) = -54$

47.
$$a - c = (-4) - (-8)$$
$$= -4 + 8$$
$$= 4$$

49. $\dfrac{b}{a} = \dfrac{6}{-4} = -\dfrac{3}{2}$

51. $5ac = 5(-4)(-8) = (-20)(-8) = 160$

53.   a.   A negative balance of trade means that imports were greater than exports.

       b. $-81.1 + (-30.7) + (-35.7) + (-68.9) + (-97) + (-95.9) + (-102.1) + (-104.7) + (-166.9) + (-265) = -\$1048$ billion

       c.

Year	Increase in Trade Balance
1991	$-30.7 - (-81.1) = 50.4$
1992	$-35.7 - (-30.7) = -5$
1993	$-68.9 - (-35.7) = -33.2$
1994	$-97 - (-68.9) = -28.1$
1995	$-95.9 - (-97) = 1.1$
1996	$-102.1 - (-95.9) = -6.2$
1997	$-104.7 - (-102.1) = -2.6$
1998	$-166.9 - (-104.7) = -62.2$
1999	$-265 - (-166.9) = -98.1$

55.

$$m + 6 = -8$$
$$\underline{-6 \qquad -6}$$
$$m = -14$$

57.

$$-72 = -8n$$
$$\frac{-72}{-8} = \frac{-8n}{-8}$$
$$9 = n$$
$$n = 9$$

59.

$$-16 + b = 20$$
$$\underline{+16 \qquad +16}$$
$$b = 36$$

61.

$$\frac{h}{6} = -12$$
$$6\left(\frac{h}{6}\right) = (-12)6$$
$$h = -72$$

63.

$$-2.9 = 0.4d$$
$$\frac{-2.9}{0.4} = \frac{0.4d}{0.4}$$
$$-7.25 = d$$
$$d = -7.25$$

65.

$$6.7 = -4.8 + y$$
$$\underline{+4.8 \qquad +4.8}$$
$$11.5 = y$$
$$y = 11.5$$

67.   a.   Imogene's net worth last year: N
$$N + 2500 = -4800$$

       b.

$$N + 2500 = -4800$$
$$\underline{-2500 \qquad -2500}$$
$$N = -7300$$

Her net worth last year was $-\$7300$.

69.   a.   Number of Super Sodas sold: n
$$-0.45n = -542.70$$

       b.

$$-0.45n = -542.70$$
$$\frac{-0.45n}{-0.45} = \frac{-542.70}{-0.45}$$
$$n = 1206$$

They sold 1206 Super Sodas.

71.   a.   Starting price of GenVec stock: s

$s - 2.50 = 7.50$

$\underline{+2.50 \quad +2.50}$

$s = 10$

The starting price was \$10 per share.

       b. $100(7.50) = \$750$

73.

75.   20

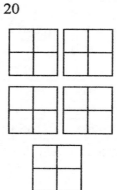

77.   a.

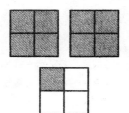

       b. $2\dfrac{1}{4} = \dfrac{9}{4}$

79.   a.

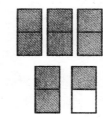

       b. $\dfrac{9}{2} = 4\dfrac{1}{2}$

81. $1 - \dfrac{2}{9} = \dfrac{9}{9} - \dfrac{2}{9} = \dfrac{7}{9}$

83. $\dfrac{2}{3}(18) = 12$

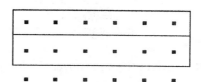

Homework 4.1 A

1. $2(4) - 3 = 8 - 3 = 5$

3. $2 + 4 \cdot 3 = 2 + 12 = 14$

5. $25 - 3(6.4) = 25 - 19.2 = 5.8$

7. $24 \div 6 + 2 = 4 + 2 = 6$

9. $80 - 56 \div 8 = 80 - 7 = 73$

11. $18 + \dfrac{18}{3} = 18 + 6 = 24$

13.
$$
\begin{array}{r}
-4 + z = 5 \\
+4 \quad +4 \\
\hline
z = 9
\end{array}
$$

15.
$$
\begin{array}{r}
x - (-8) = 3 \\
x + 8 \quad = 3 \\
-8 \quad -8 \\
\hline
x = -5
\end{array}
$$

17.
$$
\begin{array}{r}
-12 = -5 + a \\
+5 \quad +5 \\
\hline
-7 = a \\
a = -7
\end{array}
$$

19. Let S be the amount won or lost in her second game.
$$
\begin{array}{r}
S + 15 = -23 \\
-15 \quad -15 \\
\hline
S = -38
\end{array}
$$
Grace lost $38 in the second game.

21. Let N be the net change in price per share.
$$
\begin{array}{r}
2.38 + N = 1.84 \\
-2.38 \quad -2.38 \\
\hline
N = -0.54
\end{array}
$$
The net change in the price per share was −$0.54.

23.
$$
\begin{aligned}
45 - 24 \div 4(3) &= 45 - 6(3) \\
&= 45 - 18 \\
&= 27
\end{aligned}
$$

25.
$$
\begin{aligned}
84 - 2(5)(6) &= 84 - 10(6) \\
&= 84 - 60 \\
&= 24
\end{aligned}
$$

27.
$$
\begin{aligned}
18 \cdot 5 - 3 \cdot 12 &= 90 - 3 \cdot 12 \\
&= 90 - 36 \\
&= 54
\end{aligned}
$$

29.
$$
\begin{aligned}
3(8.2) - 6(2.1) &= 24.6 - 6(2.1) \\
&= 24.6 - 12.6 \\
&= 12
\end{aligned}
$$

31.
$$
\begin{aligned}
2 + 3 \cdot 8 - 6 + 3 &= 2 + 24 - 6 + 3 \\
&= 26 - 6 + 3 \\
&= 20 + 3 \\
&= 23
\end{aligned}
$$

33.
$$
\begin{aligned}
24 \div 6 + 2 \cdot 8 \div 4 &= 4 + 2 \cdot 8 \div 4 \\
&= 4 + 16 \div 4 \\
&= 4 + 4 \\
&= 8
\end{aligned}
$$

35. $-7 - 9 = -(7 + 9) = -16$

37. $8 - (-5) = 8 + 5 = 13$

39. $-6 + 15 = 15 - 6 = 9$

41. $-15 - (-15) = -15 + 15 = 0$

43. $-4 - 4 = -(4 + 4) = -8$

45.   a.   Area of blue triangle: 4 square cm
                    Area of blue trapezoid: 8 square cm
                    Total blue area: 12 square cm
       b.   Area of red trunk: 2 square cm
       c.   Area of background fabric = area of grid − area of tree = $36 - 14 = 22$ square cm

47. $37 \times 100 = 3700$ 49. $7.43 \times 1000 = 7430$

51. $8400 \div 10{,}000 = 0.84$ 53. $36 \div 100{,}000 = 0.00036$

55. $28.6 \div 100 = 0.286$

Homework 4.1 B

1.   a. $8 + 2 \cdot 5 = 8 + 10 = 18$

       b. $(8 + 2)5 = (10)5 = 50$

3.   a. $\dfrac{24}{2 + 6} = \dfrac{24}{8} = 3$

       b. $\dfrac{24}{2} + 6 = 12 + 6 = 18$

5.   a. $(9 - 4) - 3 = (5) - 3 = 2$

       b. $9 - (4 - 3) = 9 - (1) = 8$

7.   a. $6 \cdot 8 - 6 = 48 - 6 = 42$

       b. $6(8 - 6) = 6(2) = 12$

9.   a. $\dfrac{36}{6(3)} = \dfrac{36}{18} = 2$

       b. $\dfrac{36}{6}(3) = 6(3) = 18$

11.   a. $3(5 + 8)$

       b. $3 \cdot 5 + 8$

13.   a. $25 - (12 - 8)$

       b. $25 - 12 - 8$

15. $4(3 + 5) = 4(8) = 32$

17. $12 - (6 - 4) = 12 - (2) = 10$

19. $\dfrac{48}{4(6)} = \dfrac{48}{24} = 2$

21. $400 \div (100 \div 4) = 400 \div (25) = 16$

23. $\begin{aligned} 3 + 2(6 - 1) &= 3 + 2(5) \\ &= 3 + 10 \\ &= 13 \end{aligned}$

25.
$$12 - 2(1 + 3) = 12 - 2(4)$$
$$= 12 - 8$$
$$= 4$$

27.
$$6 \cdot 10 - (2 + 7) \cdot 4 = 6 \cdot 10 - (9) \cdot 4$$
$$= 60 - 36$$
$$= 24$$

29.
$$3(9 - 5) + 5(12 - 4) = 3(4) + 5(8)$$
$$= 12 + 40$$
$$= 52$$

31.
$$\frac{5 + 19}{4} = \frac{24}{4} = 6$$

33.
$$\frac{16}{20 - 12} = \frac{16}{8} = 2$$

35.
$$\frac{30 - 9}{12 - 9} = \frac{21}{3} = 7$$

37.
$$\frac{12.8 + 24.6}{3.5} = (12.8 + 24.6) \div 3.5$$
$$= 10.685 \cdots$$
$$\approx 10.69$$

39.
$$\frac{38}{79 - 24} = 38 \div (79 - 24)$$
$$= 0.690 \cdots$$
$$\approx 0.69$$

41.
$$\frac{156 - 36.7}{2.8(7.4)} = (156 - 36.7) \div (2.8 \times 7.4)$$
$$= 5.757 \cdots$$
$$\approx 5.76$$

43.
$$-4 + x = -12$$
$$\underline{+4 \qquad +4}$$
$$x = -8$$

45.
$$\frac{x}{-4} = -12$$
$$(-4)\left(\frac{x}{-4}\right) = (-12)(-4)$$
$$x = 48$$

47.
$$18 = -9 + w$$
$$\underline{+9 \quad +9}$$
$$27 = w$$
$$w = 27$$

49.
$$18 = -9w$$
$$\frac{18}{-9} = \frac{-9w}{-9}$$
$$-2 = w$$
$$w = -2$$

51.
$$59.6 \times 0.1 = 5.96$$

53.
$$17.65 \times 0.001 = 0.01765$$

55.
$$563.7 \times 0.001 = 0.5637$$

57.
$$1248 \times 0.01 = 12.48$$

59.
$$8 \times 0.0001 = 0.0008$$

Homework 4.2 A

1.

Minutes elapsed	1	3	5	6	8	10
(Calculation)	$75+30\cdot1$	$75+30\cdot3$	$75+30\cdot5$	$75+30\cdot6$	$75+30\cdot8$	$75+30\cdot10$
Oven temperature	105	165	225	255	315	375

    a.   Multiply the number of minutes elapsed times 30 and then add 75.

    b. $75 + 30m$

3.

Luisa's living expenses	2400	3000	3600	4000	4500	5000
(Calculation)	$800+\dfrac{2400}{2}$	$800+\dfrac{3000}{2}$	$800+\dfrac{3600}{2}$	$800+\dfrac{4000}{2}$	$800+\dfrac{4500}{2}$	$800+\dfrac{5000}{2}$
Parents will pay	2000	2300	2600	2800	3050	3300

    a.   Add half of Luisa's living expenses to 800.

    b. $800 + \dfrac{a}{2}$

5.

Pints Mildred kept	4	8	12	20	28
(Calculation)	$\dfrac{80-4}{4}$	$\dfrac{80-8}{4}$	$\dfrac{80-12}{4}$	$\dfrac{80-20}{4}$	$\dfrac{80-28}{4}$
Pints for each daughter	19	18	17	15	13

    a.   Subtract the number of pints Mildred kept from 80, and then divide the result by 4.

    b. $\dfrac{80-M}{4}$

7. $12 - 2t$ 9. $2t - 12$

11. $\dfrac{12}{m-3}$ 13. $\dfrac{m-3}{12}$

15. $400 + 30u$

17. $\dfrac{R - 560}{4}$

19. $2000 - 60p$

21. Length: L
Width: W
Sum of length and width: $L + W$

23. Base: b
Height: h
Product of base and height: bh

25. Gallons of alcohol: A
Volume: V

Ratio of gallons to volume: $\dfrac{A}{V}$

27. Arnold's daily workout time: t
$7t + 5$

29. Rent: r
Utilities: u
$r + u - 60$

31. Radius: r

$\dfrac{1}{2}(r + 3)$

33.
$$\begin{aligned}
4000 - 400 - 40 - 4(4) &= 4000 - 400 - 40 - 16 \\
&= 3600 - 40 - 16 \\
&= 3560 - 16 \\
&= 3544
\end{aligned}$$

35.
$$\begin{aligned}
6 + 4 \cdot 10 + 5 \cdot 100 + 3 \cdot 1000 &= 6 + 40 + 500 + 3000 \\
&= 46 + 500 + 3000 \\
&= 546 + 3000 \\
&= 3546
\end{aligned}$$

37.
$$\begin{aligned}
13.1(0.8) - 12.4 \div 6.2 &= 10.48 - 2 \\
&= 8.48
\end{aligned}$$

39.
$$\begin{aligned}
240 \div (6 + 3 \cdot 14) &= 240 \div (6 + 42) \\
&= 240 \div (48) \\
&= 5
\end{aligned}$$

41.
$$\begin{aligned}
16 - 12 \div 4 + 3(20 - 2 \cdot 7) &= 16 - 12 \div 4 + 3(20 - 14) \\
&= 16 - 12 \div 4 + 3(6) \\
&= 16 - 3 + 18 \\
&= 13 + 18 \\
&= 31
\end{aligned}$$

43.

$$-5-3\_\_\_-5$$
$$-8\ \underline{<}\ -5$$

45.

$$-4(-6)\_\_\_10$$
$$24\ \underline{>}\ 10$$

47.

$$-8-(-2)\_\_\_-8$$
$$-8+2\_\_\_-8$$
$$-6\ \underline{>}\ -8$$

49.

$$-7-7\_\_\_0$$
$$-14\ \underline{<}\ 0$$

51. $400 + 800 = 1200$
Add $4 + 12$; follow by 2 zeros

53. $1800 - 700 = 1100$
Subtract $18 - 7$; follow by 2 zeros

55. $60 + 90 = 150$
Add $6 + 9$; follow by 1 zero

57. $250 - 120 = 130$
Subtract $25 - 12$; follow by 1 zero

59. $1500 + 1700 = 3200$
Add $15 + 17$; follow by 2 zeros

Homework 4.2 B

1.

z	0	3	10
$5z + 4$	$5(0) + 4 = 0 + 4 = \mathbf{4}$	$5(3) + 4 = 15 + 4 = \mathbf{19}$	$5(10) + 4 = 50 + 4 = \mathbf{54}$

3.

b	4	6	9
$26 - 2b$	$26 - 2(4) = 26 - 8 = \mathbf{18}$	$26 - 2(6) = 26 - 12 = \mathbf{14}$	$26 - 2(9) = 26 - 18 = \mathbf{8}$

5.

h	13	7	24
$\dfrac{h-5}{4}$	$\dfrac{13-5}{4} = \dfrac{8}{4} = \mathbf{2}$	$\dfrac{7-5}{4} = \dfrac{2}{4} = \dfrac{\mathbf{1}}{\mathbf{2}}$	$\dfrac{24-5}{4} = \dfrac{\mathbf{19}}{\mathbf{4}}$

7.

d	3	6.5	0.2
$7(d + 1)$	$7(3 + 1) = 7(4) = \mathbf{28}$	$7(6.5 + 1) = 7(7.5) = \mathbf{52.5}$	$7(0.2 + 1) = 7(1.2) = \mathbf{8.4}$

9.

m	3	12	0.5
$\dfrac{m}{3+m}$	$\dfrac{3}{3+3}=\dfrac{3}{6}=\mathbf{0.5}$	$\dfrac{12}{3+12}=\dfrac{12}{15}=\mathbf{0.8}$	$\dfrac{0.5}{3+0.5}=\dfrac{0.5}{3.5}\approx\mathbf{0.14}$

11.
$$4y - x = 4(3) - 7$$
$$= 12 - 7$$
$$= 5$$

13.
$$7(s + t) = 7(12 + 28)$$
$$= 7(40)$$
$$= 280$$

15.
$$5a - 6b = 5(10) - 6(4)$$
$$= 50 - 24$$
$$= 26$$

17.
$$\frac{3w + z}{z} = \frac{3(8) + 6}{6}$$
$$= \frac{24 + 6}{6}$$
$$= \frac{30}{6}$$
$$= 5$$

19.
$$\frac{h}{g} - k = \frac{1.2}{0.6} - 0.8$$
$$= 2 - 0.8$$
$$= 1.2$$

21.
$$P = 2l + 2w$$
$$= 2(7.4) + 2(4.8)$$
$$= 14.8 + 9.6$$
$$= 24.4$$
The fringe needs to be 24.4 feet long.

23.
$$A = \frac{h}{2}(b + c)$$
$$= \frac{15}{2}(30 + 40)$$
$$= \frac{15}{2}(70)$$
$$= 7.5(70)$$
$$= 525$$
The area is 525 square feet.

25.
$$C = \frac{5F - 160}{9}$$
$$= \frac{5(98.6) - 160}{9}$$
$$= \frac{493 - 160}{9}$$
$$= \frac{333}{9}$$
$$= 37$$
Normal body temperature is 37 °C.

27.

$$A = P + \mathrm{Pr}\,t$$
$$= 50,000 + (50,000)(0.06)(5)$$
$$= 50,000 + 15,000$$
$$= 65,000$$

She has $65,000 in her account.

29. $3(-8) = -24$

31. $3 - 8 = -5$

33. $20 \div (-5) = -4$

35. $-20(-5) = 100$

37. $5 - (-20) = 5 + 20 = 25$

39. a.

Month	Jan	Feb	Mar	Apr	May	Jun	Jul	Aug	Sep	Oct	Nov	Dec
$20 - T$ (°C)	20–(-11) = **31**	20–(-9) = **29**	20–(-3) = **23**	20–2 = **18**	20 – 7 =**13**	20 – 13 = **7**	20 – 17 = **3**	20 – 16 = **4**	20 – 12 = **8**	20 – 5 = **15**	20 – (-1) = **21**	20 – (-8) = **28**

b.

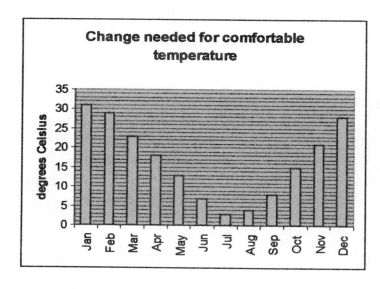

Change needed for comfortable temperature

41.

$$40 \times 80 = (4 \times 8) \text{ followed by 2 zeros}$$
$$= 3200$$

43.

$$500 \times 7000 = (5 \times 7) \text{ followed by 5 zeros}$$
$$= 3,500,000$$

45.

$$\frac{2400}{40} = (24 \div 4) \text{ followed by 1 zero}$$
$$= 60$$

47.

$$35,000 \div 700 = (35 \div 7) \text{ followed by 1 zero}$$
$$= 50$$

49.

$$160 \times 4000 = (16 \times 4) \text{ followed by 4 zeros}$$
$$= 640,000$$

Homework 4.3

1.
$$2x + 5 = 27$$
$$\underline{-5 \quad -5}$$
$$2x = 22$$
$$\frac{2x}{2} = \frac{22}{2}$$
$$x = 11$$

3.
$$4a - 6 = 14$$
$$\underline{+6 \quad +6}$$
$$4a = 20$$
$$\frac{4a}{4} = \frac{20}{4}$$
$$a = 5$$

5.
$$17 = 7t - 4$$
$$\underline{+4 \qquad +4}$$
$$21 = 7t$$
$$\frac{21}{7} = \frac{7t}{7}$$
$$3 = t$$
$$t = 3$$

7.
$$21 = 6h + 9$$
$$\underline{-9 \qquad -9}$$
$$12 = 6h$$
$$\frac{12}{6} = \frac{6h}{6}$$
$$2 = h$$
$$h = 2$$

9.
$$\frac{m}{4} - 5 = 2$$
$$\underline{+5 \qquad +5}$$
$$\frac{m}{4} = 7$$
$$4\left(\frac{m}{4}\right) = (7)4$$
$$m = 28$$

11.
$$8 + \frac{v}{3} = 10$$
$$\underline{-8 \qquad -8}$$
$$\frac{v}{3} = 2$$
$$3\left(\frac{v}{3}\right) = (2)3$$
$$v = 6$$

13.
$$\frac{4p}{5} = 8$$
$$5\left(\frac{4p}{5}\right) = (8)5$$
$$4p = 40$$
$$p = 10$$

15.
$$5(z + 4) = 35$$
$$\frac{5(z + 4)}{5} = \frac{35}{5}$$
$$z + 4 = 7$$
$$\underline{-4 \quad -4}$$
$$z = 3$$

17.

$$36 = 9(f - 7)$$

$$\frac{36}{9} = \frac{9(f - 7)}{9}$$

$$4 = f - 7$$

$$\underline{+7 \qquad +7}$$

$$11 = f$$

$$f = 11$$

19.

$$5m - 0.35 = 2.4$$

$$\underline{+0.35 \quad +0.35}$$

$$5m = 2.75$$

$$\frac{5m}{5} = \frac{2.75}{5}$$

$$m = 0.55$$

21.

$$0.08d + 55.5 = 103.5$$

$$\underline{-55.5 \quad -55.5}$$

$$0.08d = 48$$

$$\frac{0.08d}{0.08} = \frac{48}{0.08}$$

$$d = 600$$

23.

$$10.2 = \frac{w}{1.8} + 2.64$$

$$\underline{-2.64 \qquad -2.64}$$

$$7.56 = \frac{w}{1.8}$$

$$1.8(7.56) = \left(\frac{w}{1.8}\right)1.8$$

$$13.608 = w$$

$$w = 13.608$$

25. Rebate: R
Sale price: S
$S - R$

27. Field goals: g
Attempts: a
$$\frac{g}{a}$$

29. Escrow fees: F
Selling price: P
$F + 0.015P$

31. Test average: T
Homework average: H
$0.62(T + H)$

33. a.

x	2	4	6	8	10	12
$3x - 6$	$3(2) - 6$ $= 6 - 6$ $= 0$	$3(4) - 6$ $= 12 - 6$ $= 6$	$3(6) - 6$ $= 18 - 6$ $= 12$	$3(8) - 6$ $= 24 - 6$ $= 18$	$3(10) - 6$ $= 30 - 6$ $= 24$	$3(12) - 6$ $= 36 - 6$ $= 30$

b. $x = 6$

35. a.

x	4	6	8	10	12	14	16
$\dfrac{4+2x}{3}$	$\dfrac{4+2(4)}{3}$	$\dfrac{4+2(6)}{3}$	$\dfrac{4+2(8)}{3}$	$\dfrac{4+2(10)}{3}$	$\dfrac{4+2(12)}{3}$	$\dfrac{4+2(14)}{3}$	$\dfrac{4+2(16)}{3}$
	$=\dfrac{4+8}{3}$	$=\dfrac{4+12}{3}$	$=\dfrac{4+16}{3}$	$=\dfrac{4+20}{3}$	$=\dfrac{4+24}{3}$	$=\dfrac{4+28}{3}$	$=\dfrac{4+32}{3}$
	$=\dfrac{12}{3}$	$=\dfrac{\mathbf{16}}{\mathbf{3}}$	$=\dfrac{\mathbf{20}}{\mathbf{3}}$	$=\dfrac{24}{3}$	$=\dfrac{\mathbf{28}}{\mathbf{3}}$	$=\dfrac{\mathbf{32}}{\mathbf{3}}$	$=\dfrac{36}{3}$
	$=\mathbf{4}$			$=\mathbf{8}$			$=\mathbf{12}$

b. $x = 16$

37.
$$298 = \frac{1}{2}(18)h + 46$$
$$298 = 9h + 46$$
$$\underline{-46 \qquad -46}$$
$$252 = 9h$$
$$\frac{252}{9} = \frac{9h}{9}$$
$$28 = h$$
$$h = 28$$

39.
$$1080 = 600 + 600(0.04)t$$
$$1080 = 600 + 24t$$
$$\underline{-600 \; -600}$$
$$480 = 24t$$
$$\frac{480}{24} = \frac{24t}{24}$$
$$20 = t$$
$$t = 20$$

41.
$$5w - 3(6.4) = 120 + 0.2(38)$$
$$5w - 19.2 = 120 + 7.6$$
$$5w - 19.2 = 127.6$$
$$\underline{+19.2 \quad +19.2}$$
$$5w = 146.8$$
$$\frac{5w}{5} = \frac{146.8}{5}$$
$$w = 29.36$$

43.
$$\frac{2g}{7} + 12 = 16$$
$$\underline{-12 \; -12}$$
$$\frac{2g}{7} = 4$$
$$7\left(\frac{2g}{7}\right) = (4)7$$
$$2g = 28$$
$$\frac{2g}{2} = \frac{28}{2}$$
$$g = 14$$

45.

$$11 = \frac{9z}{4} - 7$$

$$\underline{+7 \qquad +7}$$

$$18 = \frac{9z}{4}$$

$$4(18) = \left(\frac{9z}{4}\right)4$$

$$72 = 9z$$

$$\frac{72}{9} = \frac{9z}{9}$$

$$8 = z$$

$$z = 8$$

47.

$$\frac{3y - 4}{5} = 4$$

$$5\left(\frac{3y - 4}{5}\right) = (4)5$$

$$3y - 4 = 20$$

$$\underline{+4 \quad +4}$$

$$3y = 24$$

$$\frac{3y}{3} = \frac{24}{3}$$

$$y = 8$$

49.

$$2(5b + 1) = 32$$

$$\frac{2(5b + 1)}{2} = \frac{32}{2}$$

$$5b + 1 = 16$$

$$\underline{-1 \quad -1}$$

$$5b = 15$$

$$\frac{5b}{5} = \frac{15}{5}$$

$$b = 3$$

51. 1 meter = 100 centimeters
57 meters = <u>5700</u> centimeters

53. 1 kilogram = 1000 grams
4.25 kilograms = <u>4250</u> grams

55. 1 liter = 1000 milliliters
8.2 liters = <u>8200</u> milliliters

57. 100 centigrams = 1 gram
78 centigrams = <u>0.78</u> grams

59. 1000 meters = 1 kilometer
5456.2 meters = <u>5.4562</u> kilometers

Homework 4.4 A

1.

$$A = lw$$
$$180 = l(12)$$
$$\frac{180}{12} = \frac{l(12)}{12}$$
$$15 = l$$

It must be 15 feet long.

3.

$$A = \frac{S}{n}$$
$$146 = \frac{S}{18}$$
$$18(146) = \left(\frac{S}{18}\right)18$$
$$2628 = S$$

The sum of their weights is 2628 pounds.

5.

$$I = \text{Pr}t$$
$$170 = (2000)r(1)$$
$$170 = 2000r$$
$$\frac{170}{2000} = \frac{2000r}{2000}$$
$$0.085 = r$$

Her investment earned 8.5%.

7.

$$P = 2l + 2w$$
$$86 = 2l + 2(18)$$
$$86 = 2l + 36$$
$$\underline{-36 \qquad -36}$$
$$50 = 2l$$
$$\frac{50}{2} = \frac{2l}{2}$$
$$25 = l$$

The length is 25 meters.

9.

$$C = d + pm$$
$$75 = 15 + 0.20m$$
$$\underline{-15 \quad -15}$$
$$60 = 0.20m$$
$$\frac{60}{0.20} = \frac{0.20m}{0.20}$$
$$300 = m$$

You drove 300 miles.

11.

$$A = P + \text{Pr}t$$
$$708 = 600 + 600r(3)$$
$$708 = 600 + 1800r$$
$$\underline{-600 \quad -600}$$
$$108 = 1800r$$
$$\frac{108}{1800} = \frac{1800r}{1800}$$
$$0.06 = r$$

The interest rate is 6%.

13.

$$A = \frac{h}{2}(b+c)$$

$$54 = \frac{6}{2}(8+c)$$

$$54 = 3(8+c)$$

$$\frac{54}{3} = \frac{3(8+c)}{3}$$

$$18 = 8+c$$

$$\underline{-8 \quad -8}$$

$$10 = c$$

The other base is 10 inches long.

15. a. $12m + 9$

m	0.5	0.75	1	1.25	2
$12m + 9$	$12(0.5) + 9$ $= 6 + 9$ $= 15$	$12(0.75) + 9$ $= 9 + 9$ $= 18$	$12(1) + 9$ $= 12 + 9$ $= 21$	$12(1.25) + 9$ $= 15 + 9$ $= 24$	$12(2) + 9$ $= 24 + 9$ $= 33$

b. $12m + 9 = 24$; $m = 1.25$ inches

17. a. $\dfrac{s+156}{3}$

s	82	84	86	88	90
$\dfrac{s+156}{3}$	$\dfrac{82+156}{3}$ $=\dfrac{238}{3}$ $= 79.\overline{3}$	$\dfrac{84+156}{3}$ $=\dfrac{240}{3}$ $= 80$	$\dfrac{86+156}{3}$ $=\dfrac{242}{3}$ $= 80.\overline{6}$	$\dfrac{88+156}{3}$ $=\dfrac{244}{3}$ $= 81.\overline{3}$	$\dfrac{90+156}{3}$ $=\dfrac{246}{3}$ $= 82$

b. $\dfrac{s+156}{3} = 80$; $s = 84$ points

19.

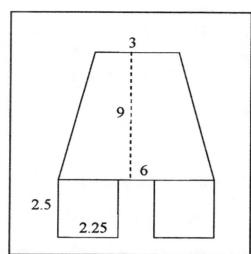

Red area = area of trapezoid + area of rectangles

$$= \frac{1}{2}h(b+c)+2lw$$

$$= \frac{1}{2}(9)(3+6)+2(2.5)(2.25)$$

$$= 40.5+11.25$$

$$= 51.75 \text{ square inches}$$

21.
$$6(x-4)=114$$
$$\frac{6(x-4)}{6}=\frac{114}{6}$$
$$x-4=19$$
$$\underline{+4 \quad +4}$$
$$x=23$$

23.
$$\frac{9x+2}{5}=13$$
$$5\left(\frac{9x+2}{5}\right)=(13)5$$
$$9x+2=65$$
$$\underline{-2 \quad -2}$$
$$9x=63$$
$$\frac{9x}{9}=\frac{63}{9}$$
$$x=7$$

25.
$$\frac{2}{3}(x+8)=12$$
$$3\left(\frac{2}{3}(x+8)\right)=(12)3$$
$$2(x+8)=36$$
$$\frac{2(x+8)}{2}=\frac{36}{2}$$
$$x+8=18$$
$$\underline{-8 \quad -8}$$
$$x=10$$

27. $18-4(3)=18-12=6$

29. $5\cdot6-2=30-2=28$

31. $3\cdot12-2\cdot4=36-8=28$

33.
$$3(12 - 2 \cdot 4) = 3(12 - 8)$$
$$= 3(4)$$
$$= 12$$

35.
$$35 - (3 \cdot 15 - 2 \cdot 6) = 35 - (45 - 12)$$
$$= 35 - (33)$$
$$= 2$$

37. $-17 - (-12) = -17 + 12 = -5$

39. $-15 - 6 = -21$

41. $9(-11) = -99$

43. $8 - (-9) = 8 + 9 = 17$

45. $-5(-5) = 25$

47. $6 \times 53 \approx 6 \times 50 = 300$

49. $4 \times 865 \approx 4 \times 900 = 3600$

51. $1493 \div 5 \approx 1500 \div 5 = 300$

53. $\dfrac{21{,}440}{7} \approx \dfrac{21{,}000}{7} = 3{,}000$

55. $8 \times 29{,}874 \approx 8 \times 30{,}00 = 240{,}000$

Homework 4.4 B

1.
$$3y - 12 = 30$$
$$\underline{+12 \quad +12}$$
$$3y = 42$$
$$\frac{3y}{3} = \frac{42}{3}$$
$$y = 14$$
There were 14 cards in each box.

3.
$$\frac{y}{3} - 12 = 30$$
$$\underline{+12 \quad +12}$$
$$\frac{y}{3} = 42$$
$$3\left(\frac{y}{3}\right) = (42)3$$
$$y = 126$$
It is 126 miles to the park.

5.
$$4x + 20 = 80$$
$$\underline{-20 \quad -20}$$
$$4x = 60$$
$$\frac{4x}{4} = \frac{60}{4}$$
$$x = 15$$
Each compact disk was $15.

7.
$$\frac{x}{4} + 20 = 80$$
$$\underline{-20 \quad -20}$$
$$\frac{x}{4} = 60$$
$$4\left(\frac{x}{4}\right) = (60)4$$
$$x = 240$$
The ticket won $240.

9. a. Number of tapes bought: n
    b. Write two expressions for the total cost of Irwin's order:
        $7n + 2 = 30$
    c.
        $$7n + 2 = 30$$
        $$\underline{-2 \quad -2}$$
        $$7n = 28$$
        $$\frac{7n}{7} = \frac{28}{7}$$
        $$n = 4$$
        Irwin bought 4 tapes.

11. a. Hours Fran must work: n
    b. Write two expressions for the amount Fran will have:
        $4n + 30 = 58$
    c.
        $$4n + 30 = 58$$
        $$\underline{-30 \quad -30}$$
        $$4n = 28$$
        $$\frac{4n}{4} = \frac{28}{4}$$
        $$n = 7$$
        Fran must work 7 hours.

13. a. Cost of the camera: c
    b. Write two expressions for the cost of the stereo:
        $4c + 9 = 269$
    c.
        $$4c + 9 = 269$$
        $$\underline{-9 \quad -9}$$
        $$4c = 260$$
        $$\frac{4c}{4} = \frac{260}{4}$$
        $$c = 65$$
        The camera costs $65.

15. a. Georgia's adjusted income: A
    b. Write two expressions for Georgia's state taxes:
        $600 + 0.07A = 2840$
    c.
        $$600 + 0.07A = 2840$$
        $$\underline{-600 \qquad -600}$$
        $$0.07A = 2240$$
        $$\frac{0.07A}{0.07} = \frac{2240}{0.07}$$
        $$A = 32{,}000$$
        Georgia's adjusted income was $32,000.

17. $-12 - 8 + 6 = -20 + 6 = -14$

19.
    $$36 - (-20) - 16 = 36 + 20 - 16$$
    $$= 56 - 16$$
    $$= 40$$

21. $48 \div (-4)(-2) = (-12)(-2) = 24$

23. $-100 \div 20 \div (-5) = -5 \div (-5) = 1$

25.
    $$-3(-3)(-3)(-3) = 9(-3)(-3)$$
    $$= (-27)(-3)$$
    $$= 81$$

27.   a.

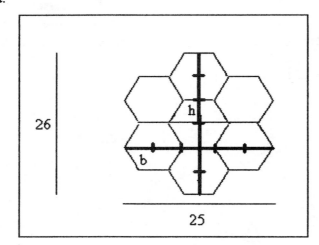

$$h = \frac{26}{6} = 4.\overline{3}$$

$$b = \frac{25}{5} = 5$$

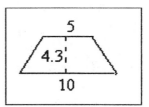

Short base: 5 cm
Long base: 10 cm
Height: $4\frac{1}{3}$ cm

b.

Blue area = 12(area of one trapezoid)

$$= 12\left(\frac{1}{2}(4.\overline{3})(5+10)\right)$$

$$= 12(32.5)$$

$$= 390 \text{ square cm}$$

29. $758 + 635 \approx 800 + 600 = 1400$

31. $2378 + 1327 \approx 2400 + 1300 = 3700$

33. $883 - 504 \approx 900 - 500 = 400$

35. $43 + 28 \approx 40 + 30 = 70$

37. $256 - 138 \approx 260 - 140 = 120$

Homework 4.5 A

1.   a. $6 + 2 - 12 + 3 = 6 + 2 + (-12) + 3$
      b.

   $$6 + 2 - 12 + 3 = 6 + 2 + (-12) + 3$$
   $$= 8 + (-12) + 3$$
   $$= -4 + 3$$
   $$= -1$$

3.   a. $-3 - 4 - (-2) = -3 + (-4) + 2$
      b.

   $$-3 - 4 - (-2) = -3 + (-4) + 2$$
   $$= -7 + 2$$
   $$= -5$$

5.   a. $6 - 8 - (-9) - 1 = 6 + (-8) + 9 + (-1)$
      b.

   $$6 - 8 - (-9) - 1 = 6 + (-8) + 9 + (-1)$$
   $$= -2 + 9 + (-1)$$
   $$= 7 + (-1)$$
   $$= 6$$

7.   a. $-28 - (-35) - 63 = -28 + 35 + (-63)$
      b.

   $$-28 - (-35) - 63 = -28 + 35 + (-63)$$
   $$= 7 + (-63)$$
   $$= -56$$

9.   a. $-12.8 + 21.7 - 19.2 + 17.5 = -12.8 + 21.7 + (-19.2) + 17.5$
      b.

   $$-12.8 + 21.7 - 19.2 + 17.5 = -12.8 + 21.7 + (-19.2) + 17.5$$
   $$= 8.9 + (-19.2) + 17.5$$
   $$= -10.3 + 17.5$$
   $$= 7.2$$

11.   a. $12 - 20 = -8$
       b. $12(-20) = -240$
       c. $12 - (-20) = 12 + 20 = 32$

13.   a. $4 - 6 - 8 = -2 - 8 = -10$
       b. $4(-6)(-8) = -24(-8) = 192$

15.   a. $-5 + 9(-16) = -5 + (-144) = -149$

       b. $-5(9) - 16 = -45 - 16 = -61$

17.   a.

$$3 - 3 - 3 + 3 = 0 - 3 + 3$$
$$= -3 + 3$$
$$= 0$$

       b.

$$3 - 3(-3)(3) = 3 - 3(-9)$$
$$= 3 - (-27)$$
$$= 3 + 27$$
$$= 30$$

19.   a.   Time Alida must wait: t

       b.   Write two expressions for the temperature of the pudding: $15 + 6t = 75$

       c.

$$15 + 6t = 75$$
$$\underline{-15 \qquad -15}$$
$$6t = 60$$
$$\frac{6t}{6} = \frac{60}{6}$$
$$t = 10$$

She must wait 10 minutes.

21.   a.   Pak's old salary: s

       b.   Write two expressions for Pak's current salary:

$$\frac{s}{3} + 500 = 8500$$

       c.

$$\frac{s}{3} + 500 = 8500$$
$$\underline{-500 \quad -500}$$
$$\frac{s}{3} = 8000$$
$$3\left(\frac{s}{3}\right) = (8000)3$$
$$s = 24{,}000$$

His old salary was \$24,000.

23.   a.   Two terms
$$\underline{3} - 9(-2)$$

       b.

$$3 - 9(-2) = 3 - (-18)$$
$$= 3 + 18$$
$$= 21$$

25.   a.   One term
$$\underline{-4(-2)(-3)}$$

       b. $-4(-2)(-3) = (8)(-3) = -24$

27.   a.   Two terms
$$\underline{-2(-4) + 3(-6)}$$

       b.

$$-2(-4) + 3(-6) = 8 + (-18)$$
$$= -10$$

29. a. Four terms
$$\underline{12} - \underline{8(-3)} + \underline{5} - \underline{8}$$

b.

$$12 - 8(-3) + 5 - 8 = 12 - (-24) + 5 - 8$$
$$= 12 + 24 + 5 - 8$$
$$= 36 + 5 - 8$$
$$= 41 - 8$$
$$= 33$$

31. a. Three terms
$$\underline{-240} + \underline{8 \cdot 20} - \underline{4 \cdot 10 \cdot 5}$$

b.

$$-240 + 8 \cdot 20 - 4 \cdot 10 \cdot 5 = -240 + 160 - 200$$
$$= -80 - 200$$
$$= -280$$

33. a. Four terms
$$\underline{-5} - \underline{2(-4)(9)} + \underline{7} - \underline{2}$$

b.

$$-5 - 2(-4)(9) + 7 - 2 = -5 - (-72) + 7 - 2$$
$$= -5 + 72 + 7 - 2$$
$$= 67 + 7 - 2$$
$$= 74 - 2$$
$$= 72$$

35. $-120 \div 6(-2) = (-20)(-2) = 40$

37.
$$90 \div (-6) \div (-3) \cdot 2 = (-15) \div (-3) \cdot 2$$
$$= 5 \cdot 2$$
$$= 10$$

39.
$$-10(-18) \div 6(-3) = 180 \div 6(-3)$$
$$= 30(-3)$$
$$= -90$$

41. $-72 - 51 \approx -70 - 50 = -120$

43. $-48 + 67 \approx -50 + 70 = 20$

45. $-77 + 43 \approx -80 + 40 = -40$

47. $-567 - 583 \approx -600 - 600 = -1200$

49. $-739 + 273 \approx -700 + 300 = -400$

Homework 4.5 B

1. a. $4 - 7 = -3$
    b. $4(-7) = -28$
    c. $4 - (-7) = 4 + 7 = 11$
    d. $-(4 - 7) = -(-3) = 3$

3. $-8 - 3(-2) = -8 - (-6) = -8 + 6 = -2$

5. $(-8)(-3)(-2) = 24(-2) = -48$

7. $-8(-3 - 2) = -8(-5) = 40$

9. $-8 - 3 - 2 = -11 - 2 = -13$

11. $-(8 - 3 - 2) = -(5 - 2) = -3$

13. $8 - (-3 - 2) = 8 - (-5) = 8 + 5 = 13$

15.
$$10 - (20 - 30) = 10 - (-10)$$
$$= 10 + 10$$
$$= 20$$

17. $-120 \div (30 \div 6) = -120 \div (5) = -24$

19. $16[20 \div (-5)] = 16[-4] = -64$

21.
$$-9 + 12 \div (-3)(8 - 2) = -9 + 12 \div (-3)(6)$$
$$= -9 + (-4)(6)$$
$$= -9 + (-24)$$
$$= -33$$

23.
$$-10 - (6 - 9) + 2(8 - 13) = -10 - (-3) + 2(-5)$$
$$= -10 - (-3) + (-10)$$
$$= -10 + 3 + (-10)$$
$$= -7 + (-10)$$
$$= -17$$

25. $\dfrac{36}{6 - 10} = \dfrac{36}{-4} = -9$

27. $\dfrac{-30 + 15}{3 - 8} = \dfrac{-15}{-5} = 3$

29.
$$\frac{8 - 2(-17)}{4 \cdot 6 - 3} = \frac{8 - (-34)}{24 - 3}$$
$$= \frac{8 + 34}{21}$$
$$= \frac{42}{21}$$
$$= 2$$

31.
$$\frac{-4(15 - 33)}{6(5 - 11)} = \frac{-4(-18)}{6(-6)}$$
$$= \frac{72}{-36}$$
$$= -2$$

33.

z	-2	0	-5
$15 - 5z$	$15 - 5(-2)$ $= 15 + 10$ $= \mathbf{25}$	$15 - 5(0)$ $= 15 + 0$ $= \mathbf{15}$	$15 - 5(-5)$ $= 15 + 25$ $= \mathbf{40}$

35.

a	-4	4	2
$-2(3a - 8)$	$-2[3(-4) - 8]$ $= -2[-12 - 8]$ $= -2[-20]$ $= \mathbf{40}$	$-2[3(4) - 8]$ $= -2[12 - 8]$ $= -2[4]$ $= \mathbf{-8}$	$-2[3(2) - 8]$ $= -2[6 - 8]$ $= -2[-2]$ $= \mathbf{4}$

37.

c	-7	-2	2
$(c + 3)(c - 4)$	$[(-7) + 3][(-7) - 4]$ $= [-4][-11]$ $= \mathbf{44}$	$[(-2) + 3][(-2) - 4]$ $= [1][-6]$ $= \mathbf{-6}$	$(2 + 3)(2 - 4)$ $= (5)(-2)$ $= \mathbf{-10}$

39.

u	-4	3	5
$\dfrac{3u}{u - 5}$	$\dfrac{3(-4)}{(-4) - 5} = \dfrac{-12}{-9}$ $= \dfrac{4}{3}$	$\dfrac{3(3)}{3 - 5} = \dfrac{9}{-2}$ $= -\dfrac{9}{2}$	$\dfrac{3(5)}{(5) - 5} = \dfrac{15}{0}$ **undefined**

41.

r	-6	0	2
$\dfrac{r - 2}{r + 2}$	$\dfrac{(-6) - 2}{(-6) + 2} = \dfrac{-8}{-4}$ $= \mathbf{2}$	$\dfrac{0 - 2}{0 + 2} = \dfrac{-2}{2}$ $= \mathbf{-1}$	$\dfrac{2 - 2}{2 + 2} = \dfrac{0}{4}$ $= \mathbf{0}$

43.
$$-3 - 4pq = -3 - 4(-3)(-5)$$
$$= -3 - (-12)(-5)$$
$$= -3 - (60)$$
$$= -63$$

45.
$$4 - b(2 + 3c) = 4 - 6[2 + 3(-8)]$$
$$= 4 - 6[2 + (-24)]$$
$$= 4 - 6[-22]$$
$$= 4 - [-132]$$
$$= 4 + 132$$
$$= 136$$

47.
$$A = \frac{1}{2}bh$$
$$720 = \frac{1}{2}(40)h$$
$$720 = 20h$$
$$\frac{720}{20} = \frac{20h}{20}$$
$$36 = h$$
The sail's height is 36 cm.

49.
$$F = 1.8C + 32$$
$$56 = 1.8C + 32$$
$$\underline{-32 \qquad -32}$$
$$24 = 1.8C$$
$$\frac{24}{1.8} = \frac{1.8C}{1.8C}$$
$$13\frac{1}{3} = C$$

The temperature is $13\frac{1}{3}$ °C.

51. $-4(678) \approx -4(700) = -2800$

53. $-5(-394) \approx -5(-400) = 2000$

55. $\dfrac{-2625}{2} \approx \dfrac{-2600}{2} = -1300$

57. $\dfrac{-4758}{-8} \approx \dfrac{-4800}{-8} = 600$

59. $\dfrac{11,748}{-6} \approx \dfrac{12,000}{-6} = -2,000$

Chapter 4 Review

1. False; Follow the order of operations.

3. True

5. False; Always start with the innermost set of parentheses and work out.

7. a. $2 + 5 \cdot (12 \div 3)$
    b. $(2 + 5) \cdot 12 \div 3$

9. a.   The parentheses mean multiplication.
    $-8 - 3(-4) = -8 - (-12) = -8 + 12 = 4$
    b.   The parentheses tell us to perform the subtraction first.
    $-8 - (3 - 4) = -8 - (-1) = -8 + 1 = -7$

11.
$$\begin{aligned} -6 + 4 - (-9) - 2 &= -2 - (-9) - 2 \\ &= -2 + 9 - 2 \\ &= 7 - 2 \\ &= 5 \end{aligned}$$

13.
$$\begin{aligned} 13 - 8(6 - 2) &= 13 - 8(4) \\ &= 13 - 32 \\ &= -19 \end{aligned}$$

15.
$$\begin{aligned} -3 + 8(-6 - 3) &= -3 + 8(-9) \\ &= -3 + (-72) \\ &= -75 \end{aligned}$$

17. $48 \div (-6) \div (-2) = (-8) \div (-2) = 4$

19. $\dfrac{-8 + 12}{-4 + 2} = \dfrac{4}{-2} = -2$

21.
$$\begin{aligned} mx + b &= (-3)(-4) + (-6) \\ &= 12 + (-6) \\ &= 6 \end{aligned}$$

23.
$$\begin{aligned} -a(3b - c) &= -12[3(-9) - (-4)] \\ &= -12[-27 - (-4)] \\ &= -12[-27 + 4] \\ &= -12[-23] \\ &= 276 \end{aligned}$$

25.
$$\begin{aligned} \frac{2mp}{p - 2m} &= \frac{2(3)(-18)}{(-18) - 2(3)} \\ &= \frac{6(-18)}{-18 - 6} \\ &= \frac{-108}{-24} \\ &= \frac{9}{2} \end{aligned}$$

27. $-0.5(-2.4) - 1.8(0.2) = 0.84$

29. $\dfrac{4.7 - 4(3 - 7.7)}{-6.8 - 2.6} = [4.7 - 4(3 - 7.7)]/(-6.8 - 2.6) = -2.5$

31.

$$A = P + \mathrm{Pr}\,t$$
$$= 2500 + 2500(0.075)(5)$$
$$= 2500 + 937.50$$
$$= \$3437.50$$

33.

$$A = \dfrac{h}{2}(b + c)$$
$$= \dfrac{9}{2}(15 + 20)$$
$$= \dfrac{9}{2}(35)$$
$$= 4.5(35)$$
$$= 157.5 \text{ square cm}$$

35.   a. $400 + 80m$

b.
$$400 + 80(9) = 400 + 720$$
$$= \$1120$$

c.
$$400 + 80m = 1440$$
$$\underline{-400 \qquad\qquad -400}$$
$$80m = 1040$$
$$\dfrac{80m}{80} = \dfrac{1040}{80}$$
$$m = 13 \text{ months}$$

37.   a. $5(d + 0.5)$

b. $5(6 + 0.5) = 5(6.5) = 32.5 \text{ miles}$

c.
$$5(d + 0.5) = 22.5$$
$$\dfrac{5(d + 0.5)}{5} = \dfrac{22.5}{5}$$
$$d + 0.5 = 4.5$$
$$\underline{-0.5 \quad -0.5}$$
$$d = 4 \text{ miles}$$

39.   a. $\dfrac{C - 50}{10}$

b. $\dfrac{670 - 50}{10} = \dfrac{620}{10} = \$62$

c.
$$\dfrac{C - 50}{10} = 79$$
$$10\left(\dfrac{C - 50}{10}\right) = (79)10$$
$$C - 50 = 790$$
$$\underline{+50 \quad +50}$$
$$C = \$840$$

41.   List the operations performed on the variable *in order*. Then undo those operations *in reverse order*.

43.

$$5x - 8 = 16 + 2$$
$$5x - 8 = 18$$
$$\underline{+8 \quad +8}$$
$$5x = 26$$
$$\frac{5x}{5} = \frac{26}{5}$$
$$x = \frac{26}{5}$$

45.

$$57 = 3z + 16 - 4$$
$$57 = 3z + 12$$
$$\underline{-12 \qquad -12}$$
$$45 = 3z$$
$$\frac{45}{3} = \frac{3z}{3}$$
$$15 = z$$
$$z = 15$$

47.

$$2(0.7)a - 4.3 = 8.3$$
$$1.4a - 4.3 = 8.3$$
$$\underline{+4.3 \quad +4.3}$$
$$1.4a = 12.6$$
$$\frac{1.4a}{1.4} = \frac{12.6}{1.4}$$
$$a = 9$$

49.

$$3 = \frac{4h - 17}{5}$$
$$5(3) = \left(\frac{4h - 17}{5}\right)5$$
$$15 = 4h - 17$$
$$\underline{+17 \qquad +17}$$
$$32 = 4h$$
$$\frac{32}{4} = \frac{4h}{4}$$
$$8 = h$$
$$h = 8$$

51.

$$21.6 + \frac{v}{2.5} = 28.8$$
$$\underline{-21.6 \qquad -21.6}$$
$$\frac{v}{2.5} = 7.2$$
$$2.5\left(\frac{v}{2.5}\right) = (7.2)2.5$$
$$v = 18$$

53. Step 1: Number of seeds: s
Step 2: Write two expressions for
the number of saplings:
$$0.60s - 38 = 112$$
Step 3:
$$0.60s - 38 = 112$$
$$\underline{+38 \quad +38}$$
$$0.60s = 150$$
$$\frac{0.60s}{0.60} = \frac{150}{0.60}$$
$$s = 250$$
They planted 250 seeds.

55. Step 1: Regular price of kayak: p
     Step 2: Write two expressions for
           the amount of the sale:
           $8(p - 30) = 1072$
     Step 3:

     $$8(p - 30) = 1072$$
     $$\frac{8(p - 30)}{8} = \frac{1072}{8}$$
     $$p - 30 = 134$$
     $$\underline{+30 \quad +30}$$
     $$p = 164$$

     The regular price of a kayak is $164.

57. $3.57 \times 1000 = 3570$

59. $5640 \times 0.001 = 5.64$

61. $380 - 260 = 120$

63. $600 \times 300 = 180,000$

65. 1000 milliliters = 1 liter
     2350 milliliters = $\underline{2.350}$ liters

67. $228 + 593 \approx 200 + 600 = 800$

69. $4 \times 3854 \approx 4 \times 4000 = 16,000$

71. $\dfrac{-2448}{6} \approx \dfrac{-2400}{6} = -400$

73. $-718 - 798 \approx -700 - 800 = -1500$

Homework 5.1 A

1. a. $A = s^2 = 8^2 = 64$ square in.
   b. $V = s^3 = 8^3 = 512$ cubic in.

3. $V = lwh = (20)(16)(8)$
   $= 2560$ cubic cm
   2560 cubes will fit inside the box.

5. a. $A = s^2 = 20^2 = 400$ square cm
   b. The cube has 6 faces.
   $SA = 6s^2$
   $\quad = 6 \cdot 20^2$
   $\quad = 6 \cdot 400$
   $\quad = 2400$ square cm
   c. $V = s^3 = 20^3 = 8000$ cubic cm

7. An *exponent* tells us how many times to use a *base* as a factor.

9. In $(-5)^2 = 25$, -5 is the number being squared. In $-5^2 = -25$, 5 is the number being squared. The two expressions have different bases.

11. $4 \cdot 4 \cdot 4 \cdot 4 \cdot 4 \cdot 4 \cdot 4 \cdot 4 = 4^8$

13. $(0.5)(0.5)(0.5) = (0.5)^3$

15. $(-1)(-1) = (-1)^2$

17. $a \cdot a \cdot a \cdot a \cdot a \cdot a = a^6$

19. $2 \cdot 2 \cdot 2 \cdot 2q \cdot q \cdot q = 2^4 q^3$

21. a.

1^2	2^2	3^2	4^2	5^2	6^2	7^2	8^2	9^2	10^2
1	4	9	16	25	36	49	64	81	100

b.

1^3	2^3	3^3	4^3	5^3	6^3	7^3	8^3	9^3	10^3
1	8	27	64	125	216	343	512	729	1000

23. a.
   $10^1 = 10$ $\quad$ $10^2 = 100$
   $10^3 = 1000$ $\quad$ $10^4 = 10,000$
   $10^5 = 100,000$

   b. One followed by 24 zeros.

25. a. $2^3 = 2\cdot2\cdot2 = 8$
     b. $5^2 = 5\cdot5 = 25$
     c. $7^2 = 7\cdot7 = 49$
     d. $6^1 = 6$

27. a. $2^4 = 2\cdot2\cdot2\cdot2 = 16$
     b. $2^1 = 2$
     c. $4^3 = 4\cdot4\cdot4 = 64$
     d. $1^3 = 1\cdot1\cdot1 = 1$

29. a. $9^2 = 9\cdot9 = 81$
     b. $6^3 = 6\cdot6\cdot6 = 216$
     c. $10^1 = 10$
     d. $12^2 = 12\cdot12 = 144$

31. a. $(-3)^2 = (-3)\cdot(-3) = 9$
     b. $(-2)^3 = (-2)\cdot(-2)\cdot(-2) = -8$

33. a.
$$(-5)^3 = (-5)\cdot(-5)\cdot(-5)$$
$$= -125$$
     b. $(-9)^2 = (-9)\cdot(-9) = 81$

35. a. $-3^2 = -3\cdot3 = -9$
     b. $-2^3 = -2\cdot2\cdot2 = -8$

37. a. $-4^2 = -4\cdot4 = -16$
     b. $-4^3 = -4\cdot4\cdot4 = -64$

39. a. $2.6^2 = 6.76$
     b. $3.2^2 = 10.24$
     c. $4.9^3 = 117.649$
     d. $0.3^2 = 0.09$

41.
$$\frac{x}{2} + 5 = 3 - 10$$
$$\frac{x}{2} + 5 = -7$$
$$\underline{-5 \quad -5}$$
$$\frac{x}{2} = -12$$
$$2\left(\frac{x}{2}\right) = (-12)2$$
$$x = -24$$

43.
$$\frac{2x-3}{-5} = -7$$
$$(-5)\left(\frac{2x-3}{-5}\right) = (-7)(-5)$$
$$2x - 3 = 35$$
$$\underline{+3 \quad +3}$$
$$2x = 38$$
$$\frac{2x}{2} = \frac{38}{2}$$
$$x = 19$$

45.
$$2(x-3) - 5 = -7$$
$$\underline{+5 \quad +5}$$
$$2(x-3) = -2$$
$$\frac{2(x-3)}{2} = \frac{-2}{2}$$
$$x - 3 = -1$$
$$\underline{+3 \quad +3}$$
$$x = 2$$

47. a. $35 \times 100 = 3500$
     b. $35 \times 1000 = 35{,}000$
     c. $35 \times 10{,}000 = 350{,}000$

49.   a. $0.074 \times 10^2 = 0.074 \times 100 = 7.4$

     b. $0.074 \times 10^3 = 0.074 \times 1000 = 74$

     c. $0.074 \times 10^4 = 0.074 \times 10{,}000 = 740$

51. $\dfrac{3}{6} = \dfrac{1}{2}$

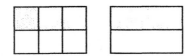

53. $\dfrac{6}{9} = \dfrac{2}{3}$

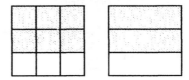

55. $\dfrac{2}{8} = \dfrac{1}{4}$

57. $\dfrac{15}{24} = \dfrac{5}{8}$

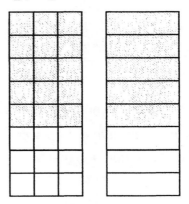

59. $\dfrac{10}{12} = \dfrac{5}{6}$

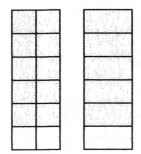

These exercises can be done without a diagram. Divide the numerator and the denominator by the greatest common factor. For example, $\dfrac{12}{15} = \dfrac{3 \cdot 4}{3 \cdot 5} = \dfrac{4}{5}$.

Homework 5.1 B

1. $6 \times 10^4 = 60,000$

3. $275 \times 10^3 = 275,000$

5. $48.623 \times 10^2 = 4862.3$

7. $7.28 \times 10^{12} = 7,280,000,000,000$

9. $0.74 \times 10^8 = 74,000,000$

11. $0.0000006 \times 10^6 = 0.6$

13.

    $113,000. \rightarrow 113.$

    Decimal point moves 3 places.

    $113,000 = 113 \times \underline{10^3}$

15.

    $5,492,000. \rightarrow 54.92$

    Decimal point moves 5 places.

    $5,492,000 = 54.92 \times \underline{10^5}$

17.

    $2,697.283 \rightarrow 2.697283$

    Decimal point moves 3 places.

    $2,697.283 = 2.697283 \times \underline{10^3}$

19.

    $643,000,000,000,000,000 \rightarrow 6.43$

    Decimal point moves 17 places.

    $643,000,000,000,000,000 = 6.43 \times \underline{10^{17}}$

21. $3,500,000 = 3.5 \times 10^6$

23. $27.6 = 2.76 \times 10^1$

25. $67,520 = 6.752 \times 10^4.$

27. $7,920,000,000,000,000,000,000 = 7.92 \times 10^{21}$

29.

    $(23,500,000,000)(187,000,000)$

    $= (2.35 \times 10^{10})(1.87 \times 10^8)$

    $= 4.3945 \times 10^{18}$

31.

    $(982,000,000,000,000) \div (4,000,000,000)$

    $= (9.82 \times 10^{14}) \div (4 \times 10^9)$

    $= 245,500 = 2.455 \times 10^5$

33.

$$\frac{(36,000,000,000)(4,800,000)}{(14,400,000,000,000)}$$

$$= \frac{(3.6 \times 10^{10})(4.8 \times 10^6)}{(14.4 \times 10^{12})}$$

$$= 12,000 = 1.2 \times 10^4$$

35.

$$\frac{(7.2 \times 10^{24})}{(1.8 \times 10^{15})(3.2 \times 10^7)}$$

$$= 125 = 1.25 \times 10^2$$

37. $(8 \times 10^{12})^3 = 5.12 \times 10^{38}$

39.

$$d = rt$$

$$2.48 \times 10^{13} = (1.86 \times 10^5)t$$

$$\frac{2.48 \times 10^{13}}{1.86 \times 10^5} = \frac{(1.86 \times 10^5)t}{1.86 \times 10^5}$$

$$1.\overline{3} \times 10^8 = t$$

$$t = 1.\overline{3} \times 10^8 \text{ sec or}$$

$$t \approx 1{,}543.2 \text{ days}$$

41.

$$D = \frac{m}{V}$$

$$= \frac{1.32 \times 10^{25}}{3.825 \times 10^{22}}$$

$$\approx 345.1 \text{ lbs/cubic ft}$$

43.   a. $3.2^4 = (3.2)(3.2)(3.2)(3.2)$

$3.2 \times 10^4 = 3.2(10)(10)(10)(10)$

In the first number, 3.2 is the base of the exponent 4.
In the second number, 10 is the base of the exponent 4.
3.2×10^4 is in scientific notation.

b. $3.2^4 = (3.2)(3.2)(3.2)(3.2) = 104.8576$

$3.2 \times 10^4 = 3.2(10)(10)(10)(10) = 32{,}000$

45.   a. $5{,}280 = 5.28 \times 10^3$

b. $365 = 3.65 \times 10^2$

c.

$$60 \cdot 60 \cdot 24 \cdot 7 = 604{,}800$$

$$= 6.048 \times 10^5$$

d. $(5{,}280)^2 \approx 2.788 \times 10^7$

47.

$$-4(-6) - 3(-4) = 24 - (-12)$$

$$= 24 + 12$$

$$= 36$$

49.

$$15 - 3(8) - 11(-2) = 15 - 24 - (-22)$$

$$= -9 + 22$$

$$= 13$$

51.

$$6 - 3[-2(-5) - 3] = 6 - 3[10 - 3]$$

$$= 6 - 3[7]$$

$$= 6 - 21$$

$$= -15$$

53.

$$(a-b)(a+b)$$
$$= [(-4)-(-3)][(-4)+(-3)]$$
$$= [-4+3][-7]$$
$$= [-1][-7]$$
$$= 7$$

55.

$$a(ab-b) = (-4)[(-4)(-3)-(-3)]$$
$$= (-4)[12+3]$$
$$= (-4)[15]$$
$$= -60$$

57. $80\% = \dfrac{80}{100} = \dfrac{4}{5}$

59. $45\% = \dfrac{45}{100} = \dfrac{9}{20}$

61. $24\% = \dfrac{24}{100} = \dfrac{6}{25}$

63. $38\% = \dfrac{38}{100} = \dfrac{19}{50}$

65. $96\% = \dfrac{96}{100} = \dfrac{24}{25}$

Homework 5.2 A

1.   a. $-4^2 = -(4)(4) = -16$

     b. $(-4)^2 = (-4)(-4) = 16$

3.   a. $6(-4)^2 = 6(16) = 96$

     b. $(6-4)^2 = (2)^2 = 4$

5.   a. $12 - 4^2 = 12 - 16 = -4$

     b. $12(-4)^2 = 12(16) = 192$

7. $4 \cdot 2^3 = 4 \cdot 8 = 32$

9. $10(-2)^4 = 10(16) = 160$

11. $(3 \cdot 4)^2 = 12^2 = 144$

13. $(2+8)^2 = (10)^2 = 100$

15. $3^2 - 6^2 = 9 - 36 = -27$

17. $3 + 5^2 = 3 + 25 = 28$

19. $3x$ means: $\underline{x + x + x}$

      x^3 means $\underline{x \cdot x \cdot x}$

21.
$$\begin{aligned} 4(2 - 3^2) &= 4(2 - 9) \\ &= 4(-7) \\ &= -28 \end{aligned}$$

23.
$$\begin{aligned} 4^2 + 8^2 \div 2^3 &= 16 + 64 \div 8 \\ &= 16 + 8 \\ &= 24 \end{aligned}$$

25.
$$\begin{aligned} (-2)(3) - 2^2 &= (-2)(3) - 4 \\ &= -6 - 4 \\ &= -10 \end{aligned}$$

27. $14 - (-3)^2 = 14 - (9) = 5$

29.
$$\begin{aligned} 5 - 4 \cdot 3^2 &= 5 - 4 \cdot 9 \\ &= 5 - 36 \\ &= -31 \end{aligned}$$

31.
$$\begin{aligned} 4 - (2 \cdot 3)^2 &= 4 - (6)^2 \\ &= 4 - 36 \\ &= -32 \end{aligned}$$

33.
$$\begin{aligned} x^4 &= (-2)^4 \\ &= (-2)(-2)(-2)(-2) \\ &= 16 \end{aligned}$$

35.
$$\begin{aligned} 2y^4 &= 2(-3)^4 \\ &= 2(81) \\ &= 162 \end{aligned}$$

37.
$$\begin{aligned} (-x)^2 &= [-(-2)]^2 \\ &= [2]^2 \\ &= 4 \end{aligned}$$

39.
$$\begin{aligned} 6 - 3x^2 &= 6 - 3(-2)^2 \\ &= 6 - 3(4) \\ &= 6 - 12 \\ &= -6 \end{aligned}$$

41.
$$-2x^3 - x = -2(-2)^3 - (-2)$$
$$= -2(-8) - (-2)$$
$$= 16 + 2$$
$$= 18$$

43.
$$2x = x + x;\ x^2 = x \cdot x$$
$$2(5) = 5 + 5 = 10;\ 5^2 = 5 \cdot 5 = 25$$

45.
$$A = lw$$
$$400 = l(16)$$
$$\frac{400}{16} = \frac{l(16)}{16}$$
$$25 = l$$
The piece is 25 feet long.

47.
$$P = rW$$
$$576 = r(1800)$$
$$\frac{576}{1800} = \frac{r(1800)}{1800}$$
$$0.32 = r$$
She spends 32% of her income on rent.

49.
$$A = \frac{S}{n}$$
$$222,960 = \frac{S}{50}$$
$$50(222,960) = \left(\frac{S}{50}\right)50$$
$$11,148,000 = S$$
There are 11,148,000 acres of state park in the U.S.

51.
$$\frac{2}{5} + \frac{4}{5} = \frac{6}{5} = 1\frac{1}{5}$$

53. $\dfrac{7}{12} + \dfrac{11}{12} = \dfrac{18}{12} = \dfrac{3}{2} = 1\dfrac{1}{2}$

55. $\dfrac{7}{9} + \dfrac{8}{9} = \dfrac{15}{9} = \dfrac{5}{3} = 1\dfrac{2}{3}$

57. $\dfrac{5}{6} + \dfrac{5}{6} = \dfrac{10}{6} = \dfrac{5}{3} = 1\dfrac{2}{3}$

59. $\dfrac{9}{16} + \dfrac{11}{16} = \dfrac{20}{16} = \dfrac{5}{4} = 1\dfrac{1}{4}$

Homework 5.2 B

1. $C = \pi d$

3. $A = \pi r^2$

5.

$$C = 2\pi r$$
$$= 2\pi(8)$$
$$= 16\pi$$
$$\approx 50.27 \text{ in.}$$

7.

$$C = 2\pi r$$
$$= 2\pi(6)$$
$$= 12\pi$$
$$\approx 37.70 \text{ ft}$$

9.

$$C = \pi d$$
$$= \pi(1.2)$$
$$\approx 3.77 \text{ m}$$

11.

$$C = \pi d$$
$$= \pi(25)$$
$$\approx 78.54 \text{ cm}$$

13.

$$C = 2\pi r$$
$$60 = 2\pi r$$
$$\frac{60}{2} = \frac{2\pi r}{2}$$
$$30 = \pi r$$
$$\frac{30}{\pi} = \frac{\pi r}{\pi}$$
$$r \approx 9.55 \text{ yd}$$

15.

$$C = 2\pi r$$
$$42.5 = 2\pi r$$
$$\frac{42.5}{2} = \frac{2\pi r}{2}$$
$$21.25 = \pi r$$
$$\frac{21.25}{\pi} = \frac{\pi r}{\pi}$$
$$r \approx 6.76 \text{ in.}$$

17.

$$A = \pi r^2$$
$$= \pi(4)^2$$
$$= 16\pi$$
$$\approx 50.27 \text{ square ft}$$

19.

$$A = \pi r^2$$
$$= \pi(0.2)^2$$
$$= 0.04\pi$$
$$\approx 0.13 \text{ square cm}$$

21.

$$A = \pi r^2 \qquad r = \frac{d}{2} = \frac{25}{2} = 12.5$$
$$= \pi(12.5)^2$$
$$= 156.25\pi$$
$$\approx 490.88 \text{ square yd}$$

23.

$$V = \frac{4}{3}\pi r^3$$
$$= \frac{4}{3}\pi(2)^3$$
$$= \frac{4}{3}\pi(8)$$
$$\approx \frac{4}{3}(25.1327)$$
$$\approx 33.51 = 3.35 \times 10^1 \text{ cubic m}$$

25.

$$V = \frac{4}{3}\pi r^3$$

$$= \frac{4}{3}\pi(0.01)^3$$

$$= \frac{4}{3}\pi(0.000001)$$

$$\approx \frac{4}{3}(3.14 \times 10^{-6})$$

$$\approx 4.19 \times 10^{-6} \text{ cubic cm}$$

27.

$$V = \frac{4}{3}\pi r^3$$

$$= \frac{4}{3}\pi(520)^3$$

$$= \frac{4}{3}\pi(1.406 \times 10^8)$$

$$\approx \frac{4}{3}(4.417 \times 10^8)$$

$$\approx 5.89 \times 10^8 \text{ cubic ft}$$

29.

$$A = \pi r^2$$

$$= \pi(7)^2$$

$$= 49\pi$$

$$\approx 153.94 \text{ square in.}$$

31.

$$C = \pi d$$

$$72.25 = \pi d$$

$$\frac{72.25}{\pi} = \frac{\pi d}{\pi}$$

$$23.00 \approx d$$

$$d \approx 23 \text{ in.}$$

33.

$$V = \frac{4}{3}\pi r^3 \quad r = \frac{d}{2} = \frac{58.7}{2} = 29.35$$

$$= \frac{4}{3}\pi(29.35)^3$$

$$= \frac{4}{3}\pi(25282.75)$$

$$\approx \frac{4}{3}(79428.10)$$

$$\approx 105{,}904.14 \text{ cubic ft}$$

35.

$$C = 2\pi r \qquad A = \pi r^2$$

$$75.4 = 2\pi r \qquad A \approx \pi(12)^2$$

$$\frac{75.4}{2\pi} = r \qquad A \approx \pi(144)$$

$$12.00 \approx r \qquad A \approx 452.39 \text{ square ft}$$

37. a. $$1{,}391{,}400 = 1.3914 \times 10^6 \; ; \; r = \frac{d}{2} = \frac{1.3914 \times 10^6}{2} = 6.957 \times 10^5 \text{ km}$$

b.

$$V = \frac{4}{3}\pi r^3$$

$$= \frac{4}{3}\pi(6.957 \times 10^5)^3$$

$$\approx \frac{4}{3}\pi(3.367 \times 10^{17})$$

$$\approx 1.41 \times 10^{18} \text{ cubic km} = 1{,}410{,}000{,}000{,}000{,}000{,}000 \text{ cubic km}$$

39.

$A = (\text{area of rectangle}) - (\text{area of semicircle})$

$\quad = (lw) - \left(\dfrac{1}{2}\pi r^2\right)$

$\quad = (10)(7) - \left(\dfrac{1}{2}\pi(3)^2\right)$

$\quad = 70 - \left(\dfrac{1}{2}\pi(9)\right)$

$\quad \approx 70 - 14.14$

$\quad = 55.86 \text{ square cm}$

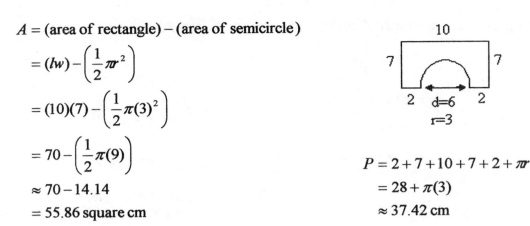

$P = 2 + 7 + 10 + 7 + 2 + \pi r$

$\quad = 28 + \pi(3)$

$\quad \approx 37.42 \text{ cm}$

41.

$A = (\text{area of rectangle}) + (\text{area of 2 semicircles}) \qquad P = 8 + 8 + (\text{circumference of 2 semicircles})$

$\quad = (lw) + \pi r^2 \qquad\qquad\qquad\qquad\qquad\qquad\qquad = 8 + 8 + 2(\pi r)$

$\quad = (8)(6) + \pi(3)^2 \qquad\qquad\qquad\qquad\qquad\qquad = 8 + 8 + 2(\pi(3))$

$\quad = 48 + 9\pi \qquad\qquad\qquad\qquad\qquad\qquad\qquad = 16 + 6\pi$

$\quad \approx 48 + 28.27 \qquad\qquad\qquad\qquad\qquad\qquad \approx 34.85 \text{ ft}$

$\quad = 76.27 \text{ square ft}$

43.

$A = (\text{area of square}) - (\text{area of circle}) \qquad P = (\text{perimeter of square}) + (\text{circumference of circle})$

$\quad = (lw) - (\pi r^2) \qquad\qquad\qquad\qquad\qquad = (12 + 12 + 12 + 12) + (2\pi r)$

$\quad = (12)(12) - (\pi(6)^2) \qquad\qquad\qquad\qquad = 48 + 2\pi(6)$

$\quad = 144 - 36\pi \qquad\qquad\qquad\qquad\qquad\quad = 48 + 12\pi$

$\quad \approx 30.90 \text{ square m} \qquad\qquad\qquad\qquad \approx 85.70 \text{ m}$

45.

Blue area $= (\text{area of 8 blocks}) - (\text{area of 8 quarter-circles})$

$\quad = 8s^2 - 8\left(\dfrac{1}{4}\pi r^2\right) \qquad\qquad r = \dfrac{4}{5}(7.5) = 6$

$\quad = 8(7.5)(7.5) - 8\left(\dfrac{1}{4}\pi(6)^2\right)$

$\quad = 450 - 72\pi$

$\quad \approx 223.8 \text{ square cm}$

47.

Area of grass reached = (area of circle) − (quarter - circle in garden)

$$= \pi r^2 - \frac{1}{4}\pi r^2$$

$$= \pi 15^2 - \frac{1}{4}\pi 15^2$$

$$= 225\pi - 56.25\pi$$

$$\approx 530.14 \text{ square ft}$$

49.

$$3^2 - (2-4)^2 = 3^2 - (-2)^2$$
$$= 9 - (4)$$
$$= 5$$

51.

$$-2 - 3(-3)^3 - 2 = -2 - 3(-27) - 2$$
$$= -2 + 81 - 2$$
$$= 79 - 2$$
$$= 77$$

53.

$$4^2 - (-4^2 - 4)^2 = 4^2 - (-16 - 4)^2$$
$$= 4^2 - (-20)^2$$
$$= 16 - (400)$$
$$= -384$$

55.

$$x^2(x - 2y) = (-3)^2[(-3) - 2(-5)]$$
$$= (-3)^2[(-3) - (-10)]$$
$$= (-3)^2[(-3) + 10]$$
$$= (-3)^2[7]$$
$$= (9)(7)$$
$$= 63$$

57.

$$-x^2 - xy^2 = -(-3)^2 - (-3)(-5)^2$$
$$= -(9) - (-3)(25)$$
$$= -9 - (-75)$$
$$= -9 + 75$$
$$= 66$$

59.

$$x - y(x^2 + y)^2 = (-3) - (-5)[(-3)^2 + (-5)]^2$$
$$= (-3) - (-5)[9 + (-5)]^2$$
$$= (-3) - (-5)[4]^2$$
$$= (-3) - (-5)[16]$$
$$= (-3) - (-80)$$
$$= -3 + 80$$
$$= 77$$

61.

$$2\frac{1}{9} + 5\frac{4}{9} = (2 + 5) + \left(\frac{1}{9} + \frac{4}{9}\right)$$
$$= 7 + \frac{5}{9}$$
$$= 7\frac{5}{9}$$

63.

$$8\frac{2}{3} - 5\frac{1}{3} = (8 - 5) + \left(\frac{2}{3} - \frac{1}{3}\right)$$
$$= 3 + \frac{1}{3}$$
$$= 3\frac{1}{3}$$

65.

$$1\frac{3}{8} + 6\frac{3}{8} = (1+6) + \left(\frac{3}{8} + \frac{3}{8}\right)$$
$$= 7 + \frac{6}{8}$$
$$= 7\frac{3}{4}$$

67.

$$9\frac{5}{6} - 4\frac{1}{6} = (9-4) + \left(\frac{5}{6} - \frac{1}{6}\right)$$
$$= 5 + \frac{4}{6}$$
$$= 5\frac{2}{3}$$

69.

$$3\frac{2}{5} + 6\frac{3}{5} = (3+6) + \left(\frac{2}{5} + \frac{3}{5}\right)$$
$$= 9 + \frac{5}{5}$$
$$= 9 + 1$$
$$= 10$$

Homework 5.3 A

1. Like terms are expressions that have the same variable factors. For example, $3x$ and $4x$ are like terms, but $3x$ and $3y$ are not.

3. False; $3x$ and $4x$ are like terms even though the coefficients are not the same.

5. True

7.
Let $y = 2$.
$5y + 3y = 5(2) + 3(2) = 10 + 6 = 16$
$8y = 8(2) = 16$

9.
Let $g = 2$.
$-9g + 2g = -9(2) + 2(2) = -18 + 4 = -14$
$-7g = -7(2) = -14$

11. $4y + 2y = (4 + 2)y = 6y$

13. $-6x + 2x = -4x$

15. $-8b + 8b = 0b = 0$

17. $-3pq + 12pq = 9pq$

19. $-32W - 47W = -79W$

21. $-7.6a - 5.2a = -12.8a$

23.
$$-12.7x - (-3.3x) = -12.7x + 3.3x$$
$$= -9.4x$$

25.
$$3x - 4x + 2x = -1x + 2x$$
$$= 1x$$
$$= x$$

27.
$$-ab + 5ab - (-3ab) = -ab + 5ab + 3ab$$
$$= 4ab + 3ab$$
$$= 7ab$$

29. $6t + 3 - 4t = 6t - 4t + 3 = 2t + 3$

31.
$$3 + 4y - (-8y) - 7 = 3 + 4y + 8y - 7$$
$$= 4y + 8y + 3 - 7$$
$$= 12y - 4$$

33.
$$-2st + 5s - 6st - (-4s) = -2st + 5s - 6st + 4s$$
$$= -2st - 6st + 5s + 4s$$
$$= -8st + 9s$$

35. $6(5d) = 30d$

37. $-8(-3h) = 24h$

39. $0.25(16a) = 4a$

41. $20(0.50m) = 10m$

43. a. The first expression is a sum.; $4x + 7x = 11x$
The second expression is a product.; $4(7x) = 28x$

b. $4x + 7x = 4(-3) + 7(-3) = -12 + (-21) = -33$
$4(7x) = 4[7(-3)] = 4[-21] = -84$

45. $25s - 3(5s) = 25s - 15s = 10s$

47. $16 - 6(4y) = 16 - 24y$

49.
$$-7(2h) - (6h - 9h) = -7(2h) - (-3h)$$
$$= -14h + 3h$$
$$= -11h$$

51.
$$-4 - 4(4t - 4t) - 4t = -4 - 4(0t) - 4t$$
$$= -4 - 0 - 4t$$
$$= -4 - 4t$$

53.
$$(5 - 9)a - 5 - 9a - 5(-9a) = -4a - 5 - 9a - (-45a)$$
$$= -4a - 5 - 9a + 45a$$
$$= -4a + 45a - 9a - 5$$
$$= 32a - 5$$

55. $8(-6)^2 = 8(36) = 288$

57.
$$12 - 8(-6)^2 = 12 - 8(36)$$
$$= 12 - 288$$
$$= -276$$

59.
$$(12 - 8) - 6^2 = (4) - 6^2$$
$$= 4 - 36$$
$$= -32$$

61.
$$-2^3(-2)^2 - 2^4 = -8(4) - 16$$
$$= -32 - 16$$
$$= -48$$

63. $\dfrac{1}{3}$ of an hour $= 20$ minutes

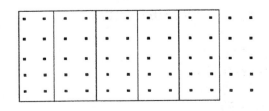

65. $\dfrac{1}{5}$ of an hour $= 12$ minutes

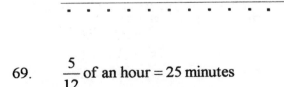

67. $\dfrac{5}{6}$ of an hour $= 50$ minutes

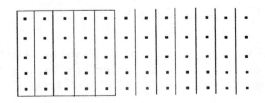

69. $\dfrac{5}{12}$ of an hour $= 25$ minutes

71. $\dfrac{8}{15}$ of an hour $= 32$ minutes

Homework 5.3 B

1. $6w^3 - 9w^3 = (6-9)w^3 = -3w^3$

3. $-8bc + 5bc = (-8+5)bc = -3bc$

5. $-pq^2 - pq^2 = (-1-1)pq^2 = -2pq^2$

7.
$$2a + 6a^2 - 3a - 5a^2$$
$$= 6a^2 - 5a^2 + 2a - 3a$$
$$= a^2 - a$$

9.
$$-4yz + 3yz^2 + 4yz + 2y^2z$$
$$= -4yz + 4yz + 3yz^2 + 2y^2z$$
$$= 3yz^2 + 2y^2z$$

11
$$5x^2 - 6x + 2$$
Cannot be simplified

13.
$$3m + m^3 + 3m^3 + 3$$
$$= m^3 + 3m^3 + 3m + 3$$
$$= 4m^3 + 3m + 3$$

15.
$$b + b^2 - ab - ab^2$$
Cannot be simplified

17. $m \cdot m^2 = m \cdot m \cdot m = m^3$

19. $k^3 \cdot k^3 = k \cdot k \cdot k \cdot k \cdot k \cdot k = k^6$

21.
$$\begin{aligned}(6x^2)(-5x^3) &= (6)(-5)x^2 \cdot x^3 \\ &= -30x \cdot x \cdot x \cdot x \cdot x \\ &= -30x^5\end{aligned}$$

23.
$$\begin{aligned}-4p(3q^2) &= (-4)(3)pq^2 \\ &= -12pq^2\end{aligned}$$

25.
$$\begin{aligned}s^3t^2(st^2) &= s^3 \cdot s \cdot t^2 \cdot t^2 \\ &= s \cdot s \cdot s \cdot s \cdot t \cdot t \cdot t \cdot t \\ &= s^4t^4\end{aligned}$$

27.

$$-6x^3z^2(-3zx^2) = (-6)(-3)x^3 \cdot x^2 \cdot z^2 \cdot z$$
$$= 18x \cdot x \cdot x \cdot x \cdot x \cdot z \cdot z \cdot z$$
$$= 18x^5z^3$$

29. a. $x + x = (1+1)x = 2x$

    b. $x \cdot x = x^2$

31. a. $x + x^2$: Cannot be simplified

    b. $x(x^2) = x(x \cdot x) = x^3$

33. a. $x^2 + x^3$: Cannot be simplified

    b. $x^2(x^3) = x \cdot x(x \cdot x \cdot x) = x^5$

35. a. $x + xy$: Cannot be simplified

    b. $x(xy) = x \cdot x \cdot y = x^2 y$

37. a. $2x + 3y$: Cannot be simplified

    b. $2x(3y) = (2)(3)xy = 6xy$

39. a. $a^2 + a^2 = (3)^2 + (3)^2 = 9 + 9 = 18$

    b. $2a^2 = 2(3)^2 = 2(9) = 18$

    c. $a^4 = (3)^4 = 81$

The expressions in (a) and (b) are equivalent since $a^2 + a^2 = 2a^2$.

41. a. $2v(3v) = 2(5)[3(5)] = 10[15] = 150$

    b. $6v = 6(5) = 30$

    c. $6v^2 = 6(5)^2 = 6(25) = 150$

The expressions in (a) and (c) are equivalent since $2v(3v) = 6v^2$.

43. a. $5w^2(-2w^2) = 5(2)^2[-2(2)^2] = 5(4)[-2(4)] = 20[-8] = -160$

    b. $3w^2 = 3(2)^2 = 3(4) = 12$

    c. $-10w^4 = -10(2)^4 = -10(16) = -160$

The expressions in (a) and (c) are equivalent since $5w^2(-2w^2) = -10w^4$.

45.

$$P = 4x + 3x + 5x$$
$$= (4 + 3 + 5)x$$
$$= 12x$$

47.

$$A = \frac{1}{2}bh$$
$$= \frac{1}{2}(8b)(15b)$$
$$= \frac{1}{2}(8)(15)b \cdot b$$
$$= 60b^2$$

49.

$$A = lw \qquad\qquad P = 2l + 2w$$
$$= (9v)(7v) \qquad = 2(9v) + 2(7v)$$
$$= 63v^2 \qquad\qquad = 18v + 14v$$
$$\qquad\qquad\qquad = 32v$$

51. $\dfrac{1}{3}\left(\dfrac{4}{5}\right) = \dfrac{4}{15}$

53. $\dfrac{3}{4}\left(\dfrac{2}{3}\right) = \dfrac{6}{12} = \dfrac{1}{2}$

55. $\dfrac{2}{3}\left(\dfrac{2}{5}\right) = \dfrac{4}{15}$

57. $\dfrac{3}{8}\left(\dfrac{1}{3}\right) = \dfrac{3}{24} = \dfrac{1}{8}$

59. $\dfrac{2}{5}\left(\dfrac{5}{8}\right) = \dfrac{10}{40} = \dfrac{1}{4}$

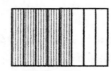

These exercises can be done without using a diagram. Multiply the numerators of the fractions to obtain the numerator in the answer and multiply the denominators of the fractions to obtain the denominator in the answer. Then simplify the fraction if necessary. For example,

$$\dfrac{3}{5}\left(\dfrac{5}{6}\right) = \dfrac{3 \cdot 5}{5 \cdot 6} = \dfrac{15}{30} = \dfrac{1}{2}.$$

Homework 5.4 A

1. a.
    6, −6
    $6^2 = 36$ and $(-6)^2 = 36$

    b.
    7, −7
    $7^2 = 49$ and $(-7)^2 = 49$

    c.
    1, −1
    $1^2 = 1$ and $(-1)^2 = 1$

3. a.
    8, −8
    $8^2 = 64$ and $(-8)^2 = 64$

    b.
    9, −9
    $9^2 = 81$ and $(-9)^2 = 81$

    c.
    12, −12
    $12^2 = 144$ and $(-12)^2 = 144$

5. When we multiply a number times itself, we get the square of the number. When we multiply the square root of a number times itself, we get the original number. For example, $4^2 = 4 \cdot 4 = 16$, but $\sqrt{4} = 2$.

7. The radical sign, $\sqrt{}$, denotes the positive square root of a number.

9. No. $\sqrt{25} \neq -5$ because the radical sign always indicates the positive square root of a number.

11. a. $\sqrt{49} = 7$
    b. $\sqrt{169} = 13$

13. a. $-\sqrt{121} = -11$
    b. $\sqrt{-25}$ is undefined

15. a. $-\sqrt{1} = -1$
    b. $\sqrt{-1}$ is undefined

17. a. $\sqrt{\dfrac{1}{16}} = \dfrac{1}{4}$
    b. $\sqrt{\dfrac{1}{25}} = \dfrac{1}{5}$

19. a. $-\sqrt{\dfrac{196}{25}} = -\dfrac{14}{5}$
    b. $\sqrt{\dfrac{4}{289}} = \dfrac{2}{17}$

21. $7\sqrt{16} = 7(4) = 28$

23.
$$3 + 5\sqrt{81} = 3 + 5(9)$$
$$= 3 + 45$$
$$= 48$$

25. $12 - \sqrt{225} = 12 - 15 = -3$

27.
$$-1-2\sqrt{9} = -1-2(3)$$
$$= -1-6$$
$$= -7$$

29.
$$\frac{6+\sqrt{144}}{3} = \frac{6+12}{3}$$
$$= \frac{18}{3}$$
$$= 6$$

31. a. $72 + 6h$

b.

$$72 + 6h = 96$$
$$\underline{-72 \qquad -72}$$
$$6h = 24$$
$$\frac{6h}{6} = \frac{24}{6}$$
$$h = 4$$

The temperature was 96° after 4 hours, at 12 noon.

33. a. $50 + \dfrac{P}{12}$

b.

$$50 + \frac{P}{12} = 68$$
$$\underline{-50 \qquad\qquad -50}$$
$$\frac{P}{12} = 18$$
$$12\left(\frac{P}{12}\right) = (18)12$$
$$P = 216$$

The party cost $216.

35. a. $\dfrac{S-60}{15}$

b.

$$\frac{S-60}{15} = 23$$
$$15\left(\frac{S-60}{15}\right) = (23)15$$
$$S - 60 = 345$$
$$\underline{+60 \quad +60}$$
$$S = 405$$

There was originally $405 in the fund.

37. a. $20 + 0.02B$

b.

$$20 + 0.02B = 340$$
$$\underline{-20 \qquad\qquad -20}$$
$$0.02B = 320$$
$$\frac{0.02B}{0.02} = \frac{320}{0.02}$$
$$B = 16{,}000$$

Her car is worth $16,000.

39. a. $-\sqrt{225} = -15$

b. $\sqrt{-225}$ cannot be simplified

41. a. $4\sqrt{25} = 4(5) = 20$

b. $\sqrt{4 \cdot 25} = \sqrt{100} = 10$

43. a.
$$5 + 2\sqrt{64} = 5 + 2(8)$$
$$= 5 + 16$$
$$= 21$$

b.
$$(5 + 2)\sqrt{64} = (7)\sqrt{64}$$
$$= (7)(8)$$
$$= 56$$

45. a. $\sqrt{100} - \sqrt{64} = 10 - 8 = 2$

b. $\sqrt{100 - 64} = \sqrt{36} = 6$

47. a.
$$10 - 3\sqrt{144} = 10 - 3(12)$$
$$= 10 - 36$$
$$= -26$$

b.
$$(10 - 3)\sqrt{144} = (7)\sqrt{144}$$
$$= (7)(12)$$
$$= 84$$

49. $3 \div \dfrac{1}{2} = 6$

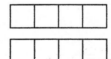

51. $3 \div \dfrac{1}{3} = 9$

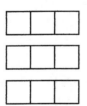

53. $2 \div \dfrac{1}{4} = 8$

55. $1 \div \dfrac{1}{8} = 8$

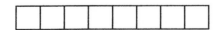

57. $4 \div \dfrac{1}{6} = 24$

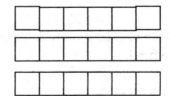

These exercises can be done without a diagram. Since all the numerators are 1, the products can be found be multiplying the whole number times the denominator of the fraction. For example,

$$5 \div \frac{1}{8} = 5 \times \frac{8}{1} = 40 .$$

Homework 5.4 B

1.

k	0	1	4
$8\sqrt{k}+12$	$8\sqrt{0}+12=8(0)+12$ $=0+12=\mathbf{12}$	$8\sqrt{1}+12=8(1)+12$ $=8+12=\mathbf{20}$	$8\sqrt{4}+12=8(2)+12$ $=16+12=\mathbf{28}$

3.

h	7	10	42
$3\sqrt{h-6}$	$3\sqrt{7-6}=3\sqrt{1}$ $=3(1)=\mathbf{3}$	$3\sqrt{10-6}=3\sqrt{4}$ $=3(2)=\mathbf{6}$	$3\sqrt{42-6}=3\sqrt{36}$ $=3(6)=\mathbf{18}$

5.

b	0	9	400
$2b-10\sqrt{b}$	$2(0)-10\sqrt{0}$ $=2(0)-10(0)$ $=0-0$ $=\mathbf{0}$	$2(9)-10\sqrt{9}$ $=2(9)-10(3)$ $=18-30$ $=\mathbf{-12}$	$2(400)-10\sqrt{400}$ $=2(400)-10(20)$ $=800-200$ $=\mathbf{600}$

7.

z	121	144	256
$(12-\sqrt{z})^2$	$(12-\sqrt{121})^2$ $=(12-11)^2$ $=(1)^2=\mathbf{1}$	$(12-\sqrt{144})^2$ $=(12-12)^2$ $=(0)^2=\mathbf{0}$	$(12-\sqrt{256})^2$ $=(12-16)^2$ $=(-4)^2=\mathbf{16}$

9.

x	36	81	225
$-3x+2x\sqrt{x}$	$-3(36)+2(36)\sqrt{36}$ $=-3(36)+2(36)(6)$ $=-108+432$ $=\mathbf{324}$	$-3(81)+2(81)\sqrt{81}$ $=-3(81)+2(81)(9)$ $=-243+1458$ $=\mathbf{1215}$	$-3(225)+2(225)\sqrt{225}$ $=-3(225)+2(225)(15)$ $=-675+6750$ $=\mathbf{6075}$

11.
$$a^2 = 25$$
$$a = \pm\sqrt{25}$$
$$a = \pm 5$$

13.
$$y^2 = 196$$
$$y = \pm\sqrt{196}$$
$$y = \pm 14$$

15.
$$3s^2 = 48$$
$$\frac{3s^2}{3} = \frac{48}{3}$$
$$s^2 = 16$$
$$s = \pm\sqrt{16}$$
$$s = \pm 4$$

17.
$$-8p^2 = -72$$
$$\frac{-8p^2}{-8} = \frac{-72}{-8}$$
$$p^2 = 9$$
$$p = \pm\sqrt{9}$$
$$p = \pm 3$$

19.
$$9m^2 = 9$$
$$\frac{9m^2}{9} = \frac{9}{9}$$
$$m^2 = 1$$
$$m = \pm\sqrt{1}$$
$$m = \pm 1$$

21.
$$7v^2 = 1183$$
$$\frac{7v^2}{7} = \frac{1183}{7}$$
$$v^2 = 169$$
$$v = \pm\sqrt{169}$$
$$v = \pm 13$$

23.
$$5x^2 - 7 = 73$$
$$\underline{+7 \quad +7}$$
$$5x^2 = 80$$
$$\frac{5x^2}{5} = \frac{80}{5}$$
$$x^2 = 16$$
$$x = \pm\sqrt{16}$$
$$x = \pm 4$$

25.
$$105 - 2w^2 = 7$$
$$\underline{-105 \qquad -105}$$
$$-2w^2 = -98$$
$$\frac{-2w^2}{-2} = \frac{-98}{-2}$$
$$w^2 = 49$$
$$w = \pm\sqrt{49}$$
$$w = \pm 7$$

27.
$$9 + \frac{Z^2}{4} = 25$$
$$\underline{-9 \qquad -9}$$
$$\frac{Z^2}{4} = 16$$
$$4\left(\frac{Z^2}{4}\right) = (16)4$$
$$Z^2 = 64$$
$$Z = \pm\sqrt{64}$$
$$Z = \pm 8$$

29.
$$\sqrt{16 + 4(12)} = \sqrt{16 + 48}$$
$$= \sqrt{64}$$
$$= 8$$

31.
$$(2+\sqrt{25})(5-\sqrt{64}) = (2+5)(5-8)$$
$$= (7)(-3)$$
$$= -21$$

33.
$$\frac{\sqrt{36}-10}{\sqrt{36}-2} = \frac{6-10}{6-2}$$
$$= \frac{-4}{4}$$
$$= -1$$

35.
$$\sqrt{4+\sqrt{25}} = \sqrt{4+5}$$
$$= \sqrt{9}$$
$$= 3$$

37. $(\sqrt{144})^2 = 144$

39. $\sqrt{9^2} = 9$

41. $\sqrt{81^2} = 81$

43. $(\sqrt{225})(\sqrt{225}) = 225$

45. a.

a	b	$a+b$	a^2	b^2	a^2+b^2	$\sqrt{a^2+b^2}$
3	4	7	9	16	25	5
5	12	17	25	144	169	13
2	6	8	4	36	40	6.3

b.   No

47. a.

a	b	$a+b$	$(a+b)^2$	a^2	b^2	a^2+b^2
2	3	5	25	4	9	13
3	5	8	64	9	25	34
4	7	11	121	16	49	65

b.   No

49. $3 \div \dfrac{3}{4} = 4$

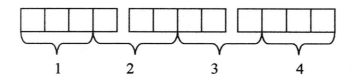

- 116 -

51. $4 \div \dfrac{4}{3} = 3$

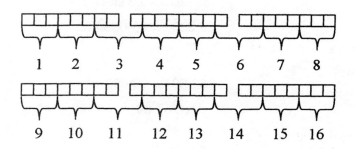

53. $6 \div \dfrac{3}{8} = 16$

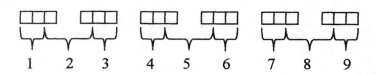

55. $6 \div \dfrac{2}{3} = 9$

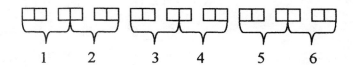

57. $9 \div \dfrac{3}{2} = 6$

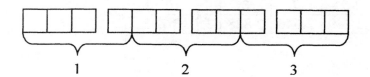

Homework 5.5 A

1.

1	2	3	4	5	6	7	8	9	10
1	1.414	1.732	2	2.236	2.449	2.646	2.828	3	3.162

3. a. $\sqrt{48.6} \approx 6.971$

  b. $\sqrt{29.3} \approx 5.413$

5. a. $\sqrt{1.4} \approx 1.183$

  b. $\sqrt{1.8} \approx 1.342$

7. a. $\sqrt{419} \approx 20.469$

  b. $\sqrt{836} \approx 28.914$

9.

$9.220 \approx \sqrt{\mathbf{85}}$ since

$9.220^2 = 85.0084$

11.

$12.961 \approx \sqrt{\mathbf{168}}$

since $12.961^2 = 167.987521$

13.

$63.891 \approx \sqrt{\mathbf{4082}}$

since $63.891^2 = 4082.059881$

15. $-2 + 6\sqrt{8} \approx 14.971$

17. $\dfrac{8 - 2\sqrt{12}}{4} \approx 0.268$

19. $\dfrac{5 - \sqrt{5}}{-3 + \sqrt{13}} \approx 4.564$

21.

$2z^2 = 10$ Check for $z = \sqrt{5}$:

$\dfrac{2z^2}{2} = \dfrac{10}{2}$ $2(\sqrt{5})^2 = 10?$

$z^2 = 5$ $2(5) = 10?$

$z = \pm\sqrt{5}$ $10 = 10$

$z \approx \pm 2.236$

23.

$3h^2 - 8 = 13$ Check for $h = \sqrt{7}$

$\underline{+8 \quad +8}$ $3(\sqrt{7})^2 - 8 = 13?$

$3h^2 = 21$ $3(7) - 8 = 13?$

$\dfrac{3h^2}{3} = \dfrac{21}{3}$ $21 - 8 = 13?$

$h^2 = 7$ $13 = 13$

$h = \pm\sqrt{7}$

$h \approx \pm 2.646$

25.

$8p^2 + 6 = 18$ Check for $p = \sqrt{3/2}$

$\underline{-6 \quad -6}$ $8\left(\sqrt{3/2}\right)^2 + 6 = 18?$

$8p^2 = 12$ $8(3/2) + 6 = 18?$

$\dfrac{8p^2}{8} = \dfrac{12}{8}$ $12 + 6 = 18?$

$p^2 = \dfrac{3}{2}$ $18 = 18$

$p = \pm\sqrt{\dfrac{3}{2}}$

$p \approx \pm 1.225$

27.

$$23.6 - 5t^2 = 4.4$$
$$\underline{-23.6 \qquad -23.6}$$
$$-5t^2 = -19.2$$
$$\frac{-5t^2}{-5} = \frac{-19.2}{-5}$$
$$t^2 = 3.84$$
$$t = \pm\sqrt{3.84}$$
$$t \approx \pm 1.960$$

Check for $t = \sqrt{3.84}$

$$23.6 - 5(\sqrt{3.84})^2 = 4.4?$$
$$23.6 - 5(3.84) = 4.4?$$
$$23.6 - 19.2 = 4.4?$$
$$4.4 = 4.4$$

29.

m	-3	2	9
$9\sqrt{m+5}$	$9\sqrt{-3+5} = 9\sqrt{2}$ $\approx \mathbf{12.728}$	$9\sqrt{2+5} = 9\sqrt{7}$ $\approx \mathbf{23.812}$	$9\sqrt{9+5} = 9\sqrt{14}$ $\approx \mathbf{33.675}$

31.

q	3	4	-6
$\sqrt{q^2-4}$	$\sqrt{(3)^2-4} = \sqrt{9-4}$ $= \sqrt{5} \approx \mathbf{2.236}$	$\sqrt{(4)^2-4} = \sqrt{16-4}$ $= \sqrt{12} \approx \mathbf{3.464}$	$\sqrt{(-6)^2-4} = \sqrt{36-4}$ $= \sqrt{32} \approx \mathbf{5.657}$

33.

a	2	5	12
$\sqrt{a} - \sqrt{a+3}$	$\sqrt{2} - \sqrt{(2)+3}$ $= \sqrt{2} - \sqrt{5} \approx \mathbf{-0.822}$	$\sqrt{5} - \sqrt{(5)+3}$ $= \sqrt{5} - \sqrt{8} \approx \mathbf{-0.592}$	$\sqrt{12} - \sqrt{(12)+3}$ $= \sqrt{12} - \sqrt{15} \approx \mathbf{-0.409}$

35.

$$m = 89.4\sqrt{h}$$
$$= 89.4\sqrt{4.7}$$
$$\approx 193.8 \text{ miles}$$

37.

$$T = 6.28\sqrt{\frac{L}{32}}$$
$$= 6.28\sqrt{\frac{75}{32}}$$
$$= 6.28\sqrt{2.34375}$$
$$\approx 9.6 \text{ seconds}$$

39.

$$A = \pi r^2 \qquad d = 2r$$
$$16 = \pi r^2 \qquad \approx 2(2.257)$$
$$\frac{16}{\pi} = r^2 \qquad \approx 4.5 \text{ feet}$$
$$\sqrt{\frac{16}{\pi}} = r$$
$$r \approx 2.257$$

41. $(\sqrt{17})^2 = 17$

43. $\sqrt{453^2} = 453$

45. $\sqrt{39}\sqrt{39} = 39$

47.

$$6x^2 + 2 = 6(\sqrt{3})^2 + 2$$
$$= 6(3) + 2$$
$$= 18 + 2$$
$$= 20$$

49.

$$-9 - 2b^2 = -9 - 2(\sqrt{11})^2$$
$$= -9 - 2(11)$$
$$= -9 - 22$$
$$= -31$$

51.

$$12 - m^2 = 12 - (\sqrt{12})^2$$
$$= 12 - (12)$$
$$= 0$$

53. The original number; For example, $\sqrt{5^2} = 5$.

55. $\dfrac{5}{8} + \dfrac{1}{8} = \dfrac{6}{8} = \dfrac{3}{4}$

57. $7\dfrac{7}{9} - 2\dfrac{4}{9} = 5\dfrac{3}{9} = 5\dfrac{1}{3}$

59. $\dfrac{8}{12} + \dfrac{6}{12} = \dfrac{14}{12} = 1\dfrac{2}{12} = 1\dfrac{1}{6}$

61. $5\dfrac{12}{16} - 3\dfrac{6}{16} = 2\dfrac{6}{16} = 2\dfrac{3}{8}$

63. $\dfrac{13}{8} - \dfrac{3}{8} = \dfrac{10}{8} = \dfrac{5}{4} = 1\dfrac{1}{4}$

Homework 5.5 B

1.

$$a^2 + b^2 = c^2$$
$$9^2 + 12^2 = c^2$$
$$81 + 144 = c^2$$
$$225 = c^2$$
$$\pm\sqrt{225} = c$$
$$\pm 15 = c$$
$$c = 15 \text{ inches}$$

3.

$$a^2 + b^2 = c^2$$
$$7^2 + b^2 = 25^2$$
$$49 + b^2 = 625$$
$$\underline{-49 \qquad -49}$$
$$b^2 = 576$$
$$b^2 = \pm\sqrt{576}$$
$$b = \pm 24$$
$$b = 24 \text{ cm}$$

5.

$$a^2 + b^2 = c^2$$
$$8^2 + 12^2 = c^2$$
$$64 + 144 = c^2$$
$$208 = c^2$$
$$\pm\sqrt{208} = c$$
$$\pm 14.422 \approx c$$
$$c \approx 14.422 \text{ miles}$$

7.

$$a^2 + b^2 = c^2$$
$$45^2 + b^2 = 62^2$$
$$2025 + b^2 = 3844$$
$$\underline{-2025 \qquad -2025}$$
$$b^2 = 1819$$
$$b^2 = \pm\sqrt{1819}$$
$$b \approx \pm 42.650$$
$$b \approx 42.650 \text{ inches}$$

9.

$$a^2 + b^2 = c^2$$
$$3.5^2 + 1.5^2 = c^2$$
$$12.25 + 2.25 = c^2$$
$$14.5 = c^2$$
$$\pm\sqrt{14.5} = c$$
$$\pm 3.808 \approx c$$
$$c \approx 3.808 \text{ feet}$$

11.

$$a^2 + b^2 = c^2$$
$$465^2 + b^2 = 820^2$$
$$216,225 + b^2 = 672,400$$
$$\underline{-216,225 \qquad -216,225}$$
$$b^2 = 456,175$$
$$b^2 = \pm\sqrt{456,175}$$
$$b \approx \pm 675.407$$
$$b \approx 675.407 \text{ mm}$$

13.

$$a^2 + b^2 = c^2$$
$$10^2 + 24^2 = c^2$$
$$100 + 576 = c^2$$
$$676 = c^2$$
$$\pm\sqrt{676} = c$$
$$\pm 26 = c$$
$$c = 26$$

He will need a ladder 26 feet long.

15.

$$a^2 + b^2 = c^2$$
$$8^2 + b^2 = 16^2$$
$$64 + b^2 = 256$$
$$\underline{-64 \qquad\quad -64}$$
$$b^2 = 192$$
$$b^2 = \pm\sqrt{192}$$
$$b \approx \pm 13.856$$
$$b \approx 13.856$$

The sand pit should be 13.856 feet from the base of the ladder.

17.

$$a^2 + b^2 = c^2$$
$$7.5^2 + b^2 = 8.5^2$$
$$56.25 + b^2 = 72.25$$
$$\underline{-56.25 \qquad\quad -56.25}$$
$$b^2 = 16$$
$$b^2 = \pm\sqrt{16}$$
$$b = \pm 4$$
$$b = 4$$

The poles should be 4 feet long.

19.

$$a^2 + b^2 = x^2 \qquad\qquad a^2 + x^2 = d^2$$
$$4^2 + 5^2 = x^2 \qquad\qquad 3^2 + 6.403^2 = d^2$$
$$16 + 25 = x^2 \qquad\qquad 9 + 41 = d^2$$
$$41 = x^2 \qquad\qquad 50 = d^2$$
$$\pm\sqrt{41} = x \qquad\qquad \pm\sqrt{50} = d$$
$$\pm 6.403 \approx x \qquad\qquad \pm 7.071 \approx d$$
$$x \approx 6.403 \qquad\qquad d \approx 7.071 \text{ feet}$$

21.   The triangle is not a right triangle. The Pythagorean Theorem applies only to right triangles.

23. a, b, and c were not squared. The equation should be $7^2 + x^2 = 9^2$.

25.   The hypotenuse is 12, not x. The equation should be $5^2 + x^2 = 12^2$.

27.   x is not squared. The equation should be $x^2 = 8^2 + 8^2$.

29.
   a. $4(5)x = 20x$

   b. $4x(5x) = (4)(5)x \cdot x = 20x^2$

   c. $4x + 5x = (4+5)x = 9x$

   d. $4 + 5x$ Cannot be simplified

31.
   a. $x^2(x^3) = x \cdot x(x \cdot x \cdot x) = x^5$

   b. $x^2 + x^3$ Cannot be simplified

   c. $x^3 + x^3 = 2x^3$

   d. $x^3(x^3) = x \cdot x \cdot x(x \cdot x \cdot x) = x^6$

33.   In $2x^3$, the base is x. In $(2x)^3$, the base is $2x$.

35.   The square of x is $x \cdot x = x^2$. The square root of x is a number whose square is x.

37.   In the first expression, the base is 1. In the second expression the base is -1. So $-1^2 = -1$ and $(-1)^2 = (-1)(-1) = 1$.

39.   In the first expression, we add first then take the square root. In the second expression, we take the square roots first and then add. So $\sqrt{9+16} = \sqrt{25} = 5$ and $\sqrt{9} + \sqrt{16} = 3 + 4 = 7$.

41. $\dfrac{3}{5}(30) = 3\left(\dfrac{30}{5}\right) = 3(6) = 18$

43. $\dfrac{3}{4}\left(\dfrac{8}{9}\right) = \dfrac{24}{36} = \dfrac{2}{3}$

45. $3 \div \dfrac{1}{3} = 3 \times 3 = 9$

47. $8 \div \dfrac{4}{5} = 8 \times \dfrac{5}{4} = \dfrac{40}{4} = 10$

49. $2 \div \dfrac{4}{3} = 2 \times \dfrac{3}{4} = \dfrac{6}{4} = \dfrac{3}{2}$

Challenge Problems

1.

$$c = \frac{15 \text{ mi} + 1 \text{ in}}{2} = \frac{15 \times 5280 \times 12 + 1}{2} = 475{,}200.5 \text{ in}$$

$$a = \frac{15 \text{ mi}}{2} = \frac{15 \times 5280 \times 12}{2} = 475{,}200 \text{ in}$$

$$a^2 + b^2 = c^2$$
$$475{,}200^2 + b^2 = 475{,}200.5^2$$
$$2.2581504 \times 10^{11} + b^2 = 2.258155152 \times 10^{11}$$
$$\underline{-2.2581504 \times 10^{11}} \qquad \underline{-2.2581504 \times 10^{11}}$$
$$b^2 = 475{,}200.25$$
$$b^2 = \pm\sqrt{475{,}200.25}$$
$$b \approx 689.348$$

c = 475,200.5 in

b?

a = 475,200 in

The rise will be approximately 689 inches or 57 feet high. Answer (d) is the closest.

Chapter 5 Review

1. A power is a base raised to an exponent. For example, 5^2 is a power.

3.

5. square inches

7. base; exponent

9. In $-4^2 = -4 \cdot 4 = -16$, the base is 4. In $(-4)^2 = (-4)(-4) = 16$, the base is -4.

11. True

13. False; Powers should be computed before multiplications.

15. True

17. False; The Pythagorean Theorem only applies to right triangles.

19. $7^3 a^5 = 7 \cdot 7 \cdot 7 a \cdot a \cdot a \cdot a \cdot a$

21. Move the decimal point in 38.762 seven places to the right.

23. The display indicates 8.6×10^{12}.

25. $5.6^3 = (5.6)(5.6)(5.6) = 175.616$ and $5.6 \times 10^3 = 5.6 \times 10 \cdot 10 \cdot 10 = 5600$

27. a. $3(-5)^2 = 3(25) = 75$
    b. $3 - 5^2 = 3 - 25 = -22$

29. a. $(-4)(-3)^2 = (-4)(9) = -36$
    b. $(-4 - 3)^2 = (-7)^2 = 49$

31. a.
$$3a^2 = 3(-4)^2$$
$$= 3(16)$$
$$= 48$$
    b.
$$(3a)^2 = [3(-4)]^2$$
$$= (-12)^2$$
$$= 144$$

33.
$$-7.4t - 5.1 + 7.4t = -7.4t + 7.4t - 5.1$$
$$= -5.1$$

35.
$$4x^2 - 3x - 2x^2 = 4x^2 - 2x^2 - 3x$$
$$= 2x^2 - 3x$$

37.   When you add like terms, the exponent remains the same. $5x^3 + 2x^3 = 7x^3$

39.   Unlike terms cannot be combined. $2c^5 - 8c^2 = 2c^5 - 8c^2$

41.   The exponents should be added, not multiplied. $p^3(p^5) = p^8$

43.   The expression cannot be simplified. $a^2b^3 = a^2b^3$

45.   a.
$$r = \frac{d}{2} = \frac{85,000}{2} = 42,500$$
$$C = 2\pi r$$
$$= 2\pi(42,500)$$
$$\approx 267,035 \text{ miles}$$

   b.
$$V = \frac{4}{3}\pi r^3$$
$$= \frac{4}{3}\pi(42,500)^3$$
$$\approx 3.2 \times 10^{14} \text{ cubic miles}$$

47.   a.
$$A = \pi r^2$$
$$= \pi(92,900,000)^2$$
$$\approx 2.7 \times 10^{16} \text{ square miles}$$

   b.
$$C = 2\pi r$$
$$= 2\pi(92,900,000)$$
$$\approx 583,707,915 \text{ miles}$$

   c.
$$d = rt$$
$$583,707,915 = 1000t$$
$$\frac{583,707915}{1000} = t$$
$$t \approx 583,708 \text{ hours}$$

   d.
$$V = \frac{4}{3}\pi r^3$$
$$= \frac{4}{3}\pi(92,900,000)^3$$
$$\approx 3.36 \times 10^{24} \text{ cubic miles}$$

49. First use the Pythagorean Theorem to find the length of the hypotenuse.

$a^2 + b^2 = c^2$

$6^2 + 6^2 = c^2$

$36 + 36 = c^2$

$72 = c^2$

$\pm\sqrt{72} = c$

$\pm 8.49 \approx c$

$c \approx 8.49$

$\text{Area} = \frac{1}{2}bh$

$= \frac{1}{2}(6)(6)$

$= 18 \text{ square feet}$

$\text{Perimeter} \approx 6 + 6 + 8.49$

$\approx 20.49 \text{ feet}$

51. First find the missing lengths.

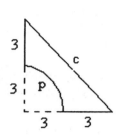

$a^2 + b^2 = c^2$

$6^2 + 6^2 = c^2$

$36 + 36 = c^2$

$72 = c^2$

$\pm\sqrt{72} = c$

$\pm 8.49 \approx c$

$c \approx 8.49$

$p = \frac{1}{4}(2\pi r)$

$= \frac{1}{4}(2\pi(3))$

$= 1.5\pi$

≈ 4.71

$\text{Area} = \text{(area of triangle)} - \text{(area of quarter - circle)}$

$= \frac{1}{2}bh - \frac{1}{4}\pi r^2$

$= \frac{1}{2}(6)(6) - \frac{1}{4}\pi(3)^2$

$= 18 - \frac{9}{4}\pi$

$\approx 10.93 \text{ square feet}$

$\text{Perimeter} \approx 3 + 3 + c + p$

$= 3 + 3 + 8.49 + 4.71$

$\approx 19.20 \text{ feet}$

53. $s^2 = (\sqrt{b})^2 = \underline{b}$

55. $\sqrt{-16}$ is undefined. $-\sqrt{16} = -4$

57. a.

$x^2 = 2.25$

$x = \sqrt{2.25} = 1.5$

b.

$\sqrt{x} = 2.25$

$x = 2.25^2 = 5.0625$

59.

$$8 - 2\sqrt{25} = 8 - 2(5)$$
$$= 8 - 10$$
$$= -2$$

61. $\dfrac{-6 + \sqrt{36 - 12}}{6} = \dfrac{-6 + \sqrt{24}}{6} \approx -0.18$

63.

$$z^2 = 81$$
$$z = \pm\sqrt{81}$$
$$z = \pm 9$$

65.

$$4w^2 - 8 = 2$$
$$\underline{+8 \quad +8}$$
$$4w^2 = 10$$
$$\frac{4w^2}{4} = \frac{10}{4}$$
$$w^2 = 2.5$$
$$w = \pm\sqrt{2.5}$$
$$w \approx \pm 1.58$$

67.

$$t = \frac{1}{7}\sqrt{\frac{h}{10}}$$
$$= \frac{1}{7}\sqrt{\frac{1000}{10}}$$
$$= \frac{1}{7}\sqrt{100}$$
$$= \frac{1}{7}(10)$$
$$= \frac{10}{7}$$
$$\approx 1.429\,\text{sec}$$

It would take the marble 1.429 seconds to fall.

69.
   a.   No; There is no 90° angle.
   b.   No; There is no 90° angle.
   c.   Yes
   d.   No; There is no 90° angle.

71. The Pythagorean Theorem states that, in a right triangle, the sum of the squares of the lengths of the legs is equal to the square of the length of the hypotenuse. If a and b are the lengths of the legs of a right triangle and c is the length of the hypotenuse, then $a^2 + b^2 = c^2$.

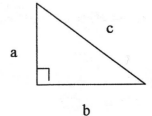

73.

$$a^2 + b^2 = c^2$$
$$6^2 + 6^2 = c^2$$
$$36 + 36 = c^2$$
$$72 = c^2$$
$$c = \sqrt{72} \text{ inches}$$

75. $\dfrac{8}{12} = \dfrac{2}{3}$

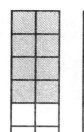

77. $28\% = \dfrac{28}{100} = \dfrac{4 \cdot 7}{4 \cdot 25} = \dfrac{7}{25}$

79. $\dfrac{1}{3}(21) = 7$

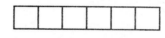

81. $\dfrac{3}{4}$ of $\dfrac{16}{9} = \dfrac{3}{4}\left(\dfrac{16}{9}\right) = \dfrac{48}{36} = \dfrac{4}{3}$

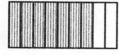

83. $6 \div \dfrac{1}{6} = 36$

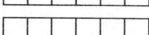
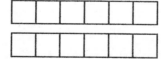

85. $10 \div \dfrac{5}{4} = 8$

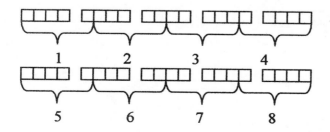

Homework 6.1

1. a. $1000
    b. March, July, November
    c. June; $1500
    d. July

3. a. 1,200,000
    b. 1973
    c. 3,800,000 + 2,100,000 =
        5,900,000
    d. 3,000,000 − 2,200,000 =
        800,000

    b. 1982 and 1989

    c. 1981 and 1986

    d. 68.5 − 60 = 8.5%

5. a.

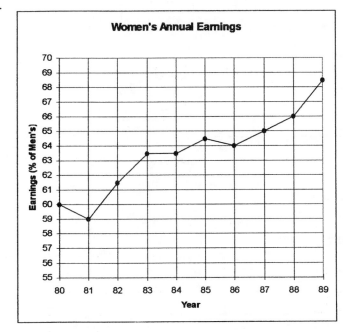

7. a.

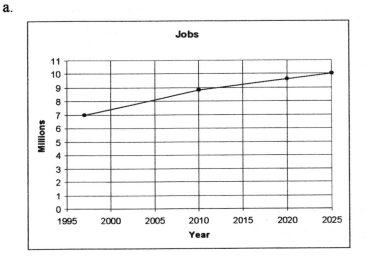

    b. The time intervals
        between bars are not
        equal.

9. a.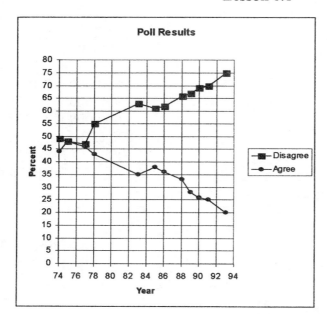

b. The percent of people who disagree with the statement is increasing, and the percent of people who agree is decreasing.

c. Some people were probably undecided.

11. a.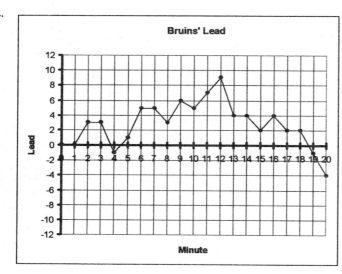

b. Minute 12

c. The Bruins were up by 9 and then down by 4. So the Cardinals gained $9 + 4 = 13$ points.

d. At the end of the first half, the Bruins were down by 4 points. The Cardinals were winning by 4 points.

13. a.
$$\frac{5-2^3}{5-2} = \frac{5-8}{3}$$
$$= \frac{-3}{3}$$
$$= -1$$

b.
$$\frac{5(-2)^3}{5(-2)} = \frac{5(-8)}{-10}$$
$$= \frac{-40}{-10}$$
$$= 4$$

15. a. $\sqrt{(4)(9)} = \sqrt{36} = 6$

b. $\left(\sqrt{4}\right)\left(\sqrt{9}\right) = (2)(3) = 6$

17.   a.
$$\sqrt{15^2 - 12^2} = \sqrt{225 - 144}$$
$$= \sqrt{81}$$
$$= 9$$

b. $\sqrt{15^2} - \sqrt{12^2} = 15 - 12 = 3$

19.   a.
$$\frac{18 + \sqrt{81}}{3} = \frac{18 + 9}{3}$$
$$= \frac{27}{3}$$
$$= 9$$

b.
$$\frac{18\sqrt{81}}{3} = \frac{18(9)}{3}$$
$$= \frac{162}{3}$$
$$= 54$$

21. $\dfrac{3}{4} = \dfrac{\mathbf{6}}{\mathbf{8}}$

23. $\dfrac{2}{3} = \dfrac{\mathbf{6}}{\mathbf{9}}$

25. $\dfrac{3}{4} = \dfrac{\mathbf{9}}{12}$

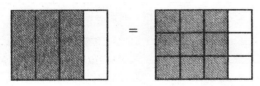

27. $\dfrac{3}{8} = \dfrac{\mathbf{6}}{16}$

29. $\dfrac{5}{8} = \dfrac{\mathbf{15}}{24}$

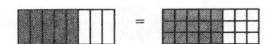

To do these exercises without a diagram, first determine by what factor the old denominator was multiplied to get the new denominator. Then multiply the numerator by that same factor. For example, in problem 30, $32 \div 8 = 4$ and $\dfrac{3}{8} = \dfrac{3 \cdot 4}{8 \cdot 4} = \dfrac{12}{32}$.

Homework 6.2 A

1.   a.   Extremes: 5 and 18
         Range: 18 – 5 – 13

     b.

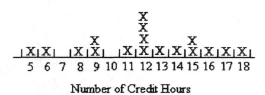

Number of Credit Hours

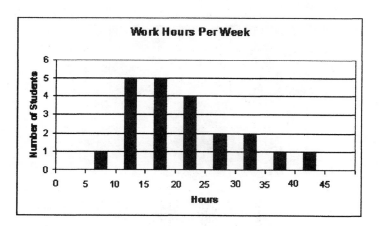

     c.

$$\text{Mean} = \frac{5+6+8+9+9+11+12+12+12+12+13+14+15+15+16+17+18}{17}$$

$$= \frac{204}{17} = 12$$

Median: 12

Mode: 12

3.   a.   Extremes: 6 and 40
         Range: 40 – 6 = 34

     b.

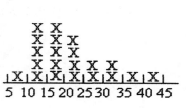

Hours per week

     c.

Mean

$$= \frac{6+10+12+12+12+12+15+15+16+18+18+20+20+22+24+25+26+30+32+35+40}{21}$$

$$= \frac{420}{21} = 20$$

Median: 18

Mode: 12

5. a. Extremes: $22,684 and $39,482
        Range: $39,482 - 22,684 = \$16,798$

    b.

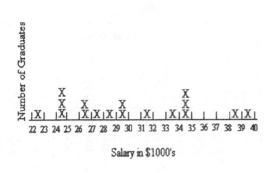

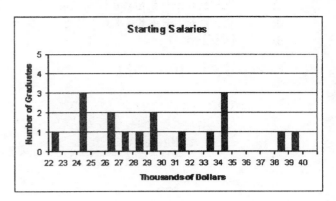

    c. Mean $= \dfrac{S}{n} = \dfrac{\$509,963}{17} \approx \$29,998$ Median: $29,211   Mode: none

7. a.

Mean $= \dfrac{100,000 + 100,000 + 110,000 + 120,000 + 170,000 + 190,000 + 2,500,000}{7}$

$= \dfrac{3,290,000}{7}$

$= \$470,000$

Median: $120,000
Mode: $100,000

    b. The mean is not the best indicator of the average value since it is severely affected by the most expensive house. The mode is also not the best indicator because it is actually the lowest value listed. The median is probably the best indicator of the average value of the houses.

    c. The median would likely be the most useful because it is closest to the typical house price.

9. a.

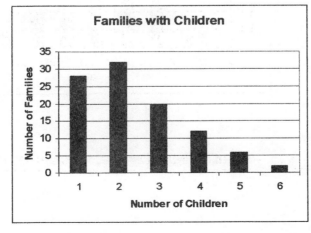

    b. Mean $= \dfrac{S}{n} = \dfrac{1 \cdot 28 + 2 \cdot 32 + 3 \cdot 20 + 4 \cdot 12 + 5 \cdot 6 + 6 \cdot 2}{28 + 32 + 20 + 12 + 6 + 2} = \dfrac{242}{100} = 2.42$

11. a. Extremes: 4 and 19; Range: $19 - 4 = 15$
    b. Mode: 11
    c. $7 + 5 + 3 + 2 + 1 = 18$ people
    d. $1 + 2 + 5 + 4 + 6 + 6 + 8 + 7 + 5 + 3 + 2 + 1 = 50$ people
    e. Median: 11
    f. $\dfrac{5}{50} = 0.10 = 10\%$
    g.
    $$\text{Mean} = \frac{S}{n}$$
    $$= \frac{4 \cdot 1 + 6 \cdot 2 + 7 \cdot 5 + 8 \cdot 4 + 9 \cdot 6 + 10 \cdot 6 + 11 \cdot 8 + 12 \cdot 7 + 13 \cdot 5 + 14 \cdot 3 + 16 \cdot 2 + 19 \cdot 1}{50}$$
    $$= \frac{527}{50} = 10.54$$

13. $\dfrac{2x}{0}$

    Division by zero is undefined.

15. $-6^3 = -216$

17. $a^2 b + 2a^2 b = 3a^2 b$

19. $2mn(3m^2 n) = 2(3)mm^2 nn = 6m^3 n^2$

21. $\sqrt{121} = 11$

23. $\dfrac{1}{3} = \dfrac{\mathbf{4}}{12}; \dfrac{3}{4} = \dfrac{\mathbf{9}}{12}$

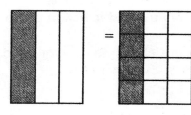

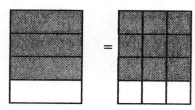

25. $\dfrac{3}{2} = \dfrac{\mathbf{15}}{10}; \dfrac{4}{5} = \dfrac{\mathbf{8}}{10}$

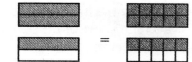

27. $\dfrac{3}{4} = \dfrac{\mathbf{9}}{12}; \dfrac{5}{6} = \dfrac{\mathbf{10}}{12}$

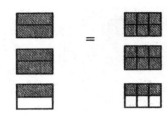

29. $\dfrac{1}{6} = \dfrac{\mathbf{4}}{24}; \dfrac{3}{8} = \dfrac{\mathbf{9}}{24}$

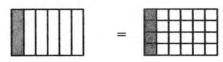

31. $\dfrac{5}{2} = \dfrac{\mathbf{15}}{6}; \dfrac{5}{3} = \dfrac{\mathbf{10}}{6}$

Homework 6.2 B

1. a. 127 135 136 142 144 145 148 148 151 152 153 156 165 166 168 174

$$\text{median} = \frac{148+151}{2} = \frac{299}{2} = 149.5 \qquad LQ = \frac{142+144}{2} = \frac{286}{2} = 143$$

$$UQ = \frac{156+165}{2} = \frac{321}{2} = 160.5$$

b. $IQR = 160.5 - 143 = 17.5$

c.

3. a.   56 58 60 64 64 65 66 67 68 68 72 74 76 78 80 82

$$\text{median} = \frac{67+68}{2} = \frac{135}{2} = 67.5 \qquad LQ = \frac{64+64}{2} = \frac{128}{2} = 64$$

$$UQ = \frac{74+76}{2} = \frac{150}{2} = 75$$

b. $IQR = 75 - 64 = 11$

c.

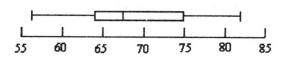

5. a.   12.5 13.6 13.9 14.0 14.0 14.1 14.2 14.5 14.5 14.6 14.6 14.6 **14.7** 14.7 14.8 14.8 15.0
        15.0 15.1 15.1 15.1 15.2 15.2 15.2 15.3 **15.3** 15.4 15.5 15.6 15.6 15.9 15.9 16.0 16.2
        16.2 16.2 16.3 16.5 **16.7** 16.7 16.8 17.0 17.2 17.6 18.0 18.2 18.3 19.7 20.1 20.4 20.5

   median $= 15.3$ $\qquad\qquad$ LQ $= 14.7$

   UQ $= 16.7$

   b. $IQR = 16.7 - 14.7 = 2$

   c.

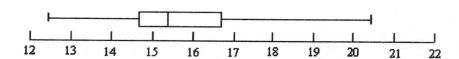

7. a.   Neighborhood A; The youngest child in A is 1 year old, and the youngest in B is 2 years old.

   b.   Neighborhood B; The oldest child in B is 17 years old, and the oldest in A is 13 years old.

   c.   Neighborhood B; The range in B is 15 years, and the range in A is 12 years.

   d.   Neighborhood A; The median age in A is 8, and the median age in B is 5.

   e.   Neighborhood B seems like the better choice. In neighborhood B, one-fourth of the children are between 3 and 5 years old. However, in neighborhood A, half the children are over 8 years old and half are between 1 and 8 years old. The median age in neighborhood B is also closer to Stefanie's age.

9. $y = -4x$ $\qquad\qquad\qquad\qquad\qquad$ 11. $q = p + 3$

13. $r = m - 2$

15. $b = \dfrac{-a}{2}$

17. $p = 3n - 1$ $\qquad\qquad\qquad\qquad\qquad$ 19. $\dfrac{6 + 2^3}{7} = \dfrac{6 + 8}{7} = \dfrac{14}{7} = 2$

21. $-\sqrt{-0} = -\sqrt{0} = -0 = 0$

23. $\sqrt{16-9} = \sqrt{7}$

25. $\dfrac{\sqrt{4}-2}{3} = \dfrac{2-2}{3} = \dfrac{0}{2} = 0$

27. $\dfrac{-w}{6-2\sqrt{9}} = \dfrac{-w}{6-2(3)} = \dfrac{-w}{6-6} = \dfrac{-w}{0}$
Division by zero is undefined.

29.

$$\dfrac{1}{4} \quad + \quad \dfrac{2}{3} \quad = \quad \dfrac{3}{12} \quad + \quad \dfrac{8}{12} \quad = \quad \dfrac{11}{12}$$

31.

$$\dfrac{3}{5} \quad + \quad \dfrac{5}{2} \quad = \quad \dfrac{6}{10} \quad + \quad \dfrac{25}{10} \quad = \quad \dfrac{31}{10}$$

33.

$$\dfrac{4}{3} \quad + \quad \dfrac{3}{2} \quad = \quad \dfrac{8}{6} \quad + \quad \dfrac{9}{6} \quad = \quad \dfrac{17}{6}$$

35.

$$\dfrac{2}{5} \quad + \quad \dfrac{3}{4} \quad = \quad \dfrac{8}{20} \quad + \quad \dfrac{15}{20} \quad = \quad \dfrac{23}{20}$$

37.

$$\dfrac{2}{3} \quad + \quad \dfrac{2}{5} \quad = \quad \dfrac{10}{15} \quad + \quad \dfrac{6}{15} \quad = \quad \dfrac{16}{15}$$

Homework 6.3 A

1.

c	d
200	1000
400	800
600	600
800	400
1000	200
1200	0

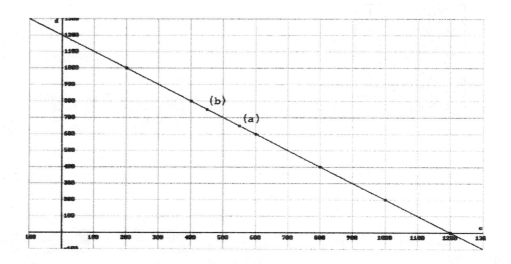

    a.   650
    b.   450

3.

r	C
2	2
3	4.5
4	8
5	12.5
6	18
7	24.5

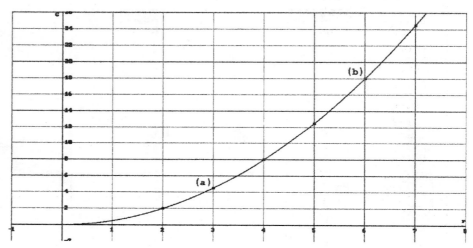

    a. $4.50
    b.   6-in radius

5.

v	t
30	20
50	12
60	10
100	6
120	5
150	4
200	3
300	2

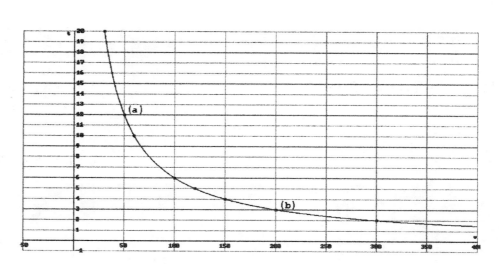

    a.   12 hours
    b.   200 mph

Lesson 6.3

7.

h	d
5000	86.3
10,000	122
15,000	149.4
20,000	172.5
25,000	192.9
30,000	211.3

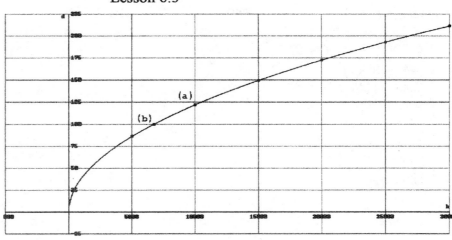

a.   122 miles
b.   6719 feet

9.   a.

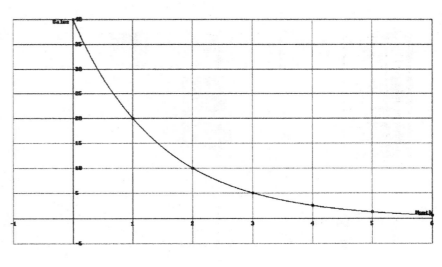

Month	Sales ($1000's)
April (m=0)	40
May (m=1)	20
June (m=2)	10
July (m=3)	5
August (m=4)	2.5
September (m=5)	1.25
October (m=6)	0.625

b.

11. $2-3(4) = 2-12 = -10$

13. $\dfrac{114-48}{-6} = \dfrac{66}{-6} = -11$

15. $\dfrac{12-4}{6-8} = \dfrac{8}{-2} = -4$

17.
$$8-3(2-5)(3-8) = 8-3(-3)(-5)$$
$$= 8-3(15)$$
$$= 8-45$$
$$= -37$$

19.

$$49 - (18 - 3 \cdot 4)(1 + 2 \cdot 3)$$
$$= 49 - (18 - 12)(1 + 6)$$
$$= 49 - (6)(7)$$
$$= 49 - 42$$
$$= 7$$

21.

$$\frac{6 - 3(5) + 9}{6 - 6(-3)} = \frac{6 - 15 + 9}{6 - (-18)}$$
$$= \frac{-9 + 9}{6 + 18}$$
$$= \frac{0}{24}$$
$$= 0$$

23. $y = -4x$

25. $q = p + 3$

27. $r = m - 2$

29. $b = \dfrac{-a}{2}$

31. $3 + \dfrac{3}{4} = \dfrac{12}{4} + \dfrac{3}{4} = \dfrac{15}{4}$

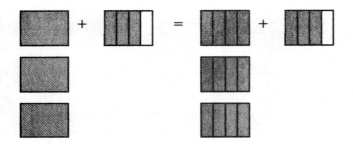

33. $2 + \dfrac{3}{8} = \dfrac{16}{8} + \dfrac{3}{8} = \dfrac{19}{8}$

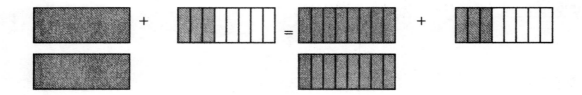

35. $4 - \dfrac{5}{6} = \dfrac{24}{6} - \dfrac{5}{6} = \dfrac{19}{6}$

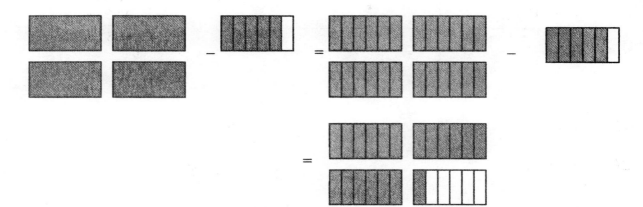

37. $2 - \dfrac{3}{5} = \dfrac{10}{5} - \dfrac{3}{5} = \dfrac{7}{5}$

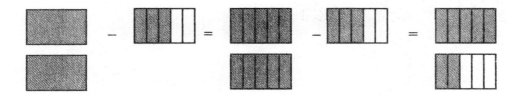

39. $1 + \dfrac{1}{3} = \dfrac{3}{3} + \dfrac{1}{3} = \dfrac{4}{3}$

Homework 6.3 B

1. The Cartesian coordinate system is formed by two axes which divide the plane into four quadrants. The x-axis is the horizontal number line, and the y-axis is the vertical number line. The origin is the point where the two axes cross and has coordinates (0, 0).

3.   a. x and y are both negative.
       b. x is negative, and y is positive.

5.   A horizontal line

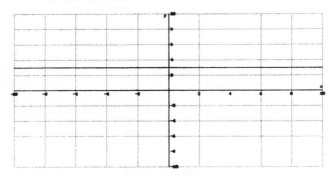

7.

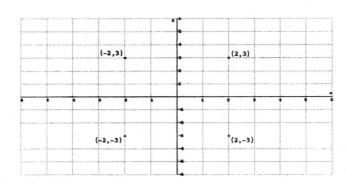

9.

11.

x	-2	-1	$\dfrac{-1}{2}$	0	$\dfrac{1}{2}$	1	2
y	-7	0	$\dfrac{7}{8}$	1	$\dfrac{9}{8}$	2	9

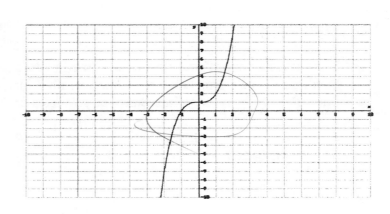

13.

x	-4	-3	-2	-1	$-\dfrac{1}{2}$	$-\dfrac{1}{4}$
y	$-\dfrac{1}{4}$	$-\dfrac{1}{3}$	$-\dfrac{1}{2}$	-1	-2	-4

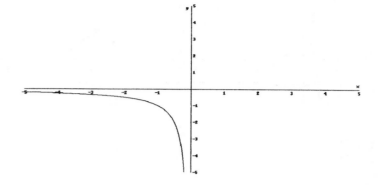

15.

x	4	5	6	7	8	13	20
y	0	1	1.4	1.7	2	3	4

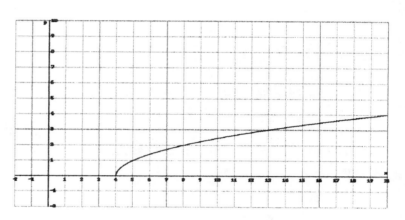

17.

x	-4	-3	-2	-1	0	1	2	3	4
y	-7	0	5	8	9	8	5	0	-7

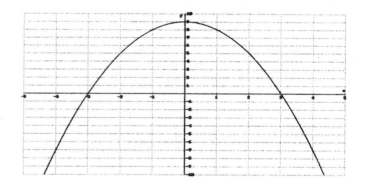

19.

x	-4	-3	-2	-1	0	1	2
y	8	3	0	-1	0	3	8

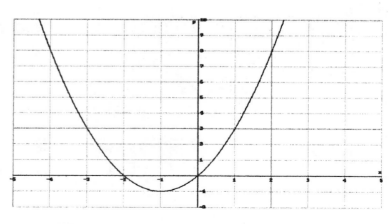

21.

x	y
-2	-1
-1	-2
$\frac{1}{2}$	4
1	2

$$y = \frac{2}{x}$$

23.

x	y
0	0
1	2
4	4
9	6

$$y = 2\sqrt{x}$$

25.

x	y
5	3
3	1
0	-2
-1	-3

$$y = x - 2$$

27.

$$\frac{x+25}{5} = 75$$

$$5\left(\frac{x+25}{5}\right) = (75)5$$

$$x + 25 = 375$$

$$\underline{-25 \quad -25}$$

$$x = 350$$

The sum of her first four exam scores is 350.

29.

$$\frac{x}{5} + 25 = 75$$

$$\underline{-25 \quad -25}$$

$$\frac{x}{5} = 50$$

$$5\left(\frac{x}{5}\right) = (50)5$$

$$x = 250$$

The food bill was $250.

31. $\dfrac{1}{3} + \dfrac{5}{6} = \dfrac{2}{6} + \dfrac{5}{6} = \dfrac{7}{6}$

33. $\dfrac{3}{4} + \dfrac{5}{16} = \dfrac{12}{16} + \dfrac{5}{16} = \dfrac{17}{16}$

     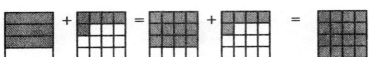

35. $\dfrac{7}{12} - \dfrac{1}{3} = \dfrac{7}{12} - \dfrac{4}{12} = \dfrac{3}{12} = \dfrac{1}{4}$

    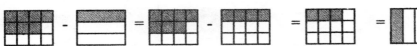

37. $\dfrac{2}{3} - \dfrac{4}{15} = \dfrac{10}{15} - \dfrac{4}{15} = \dfrac{6}{15} = \dfrac{2}{5}$

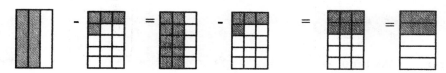

39. $\dfrac{5}{18} + \dfrac{1}{6} = \dfrac{5}{18} + \dfrac{3}{18} = \dfrac{8}{18} = \dfrac{4}{9}$

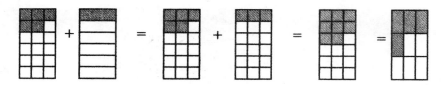

Homework 6.4

1. a. $y = 3x + 2$
    b. $y = 3x^2 + 5$

3. The x-intercept of a graph is the point at which the graph crosses the x-axis. The y-intercept of a graph is the point at which the graph crosses the y-axis.

5. a.

x	–3	3	6
y	–3	3	6

b.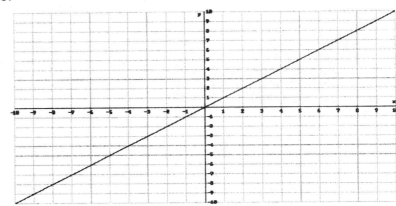

    c. x-intercept: (0, 0)
          y-intercept: (0, 0)

7. a.

x	–5	0	4
y	–6	–1	3

b.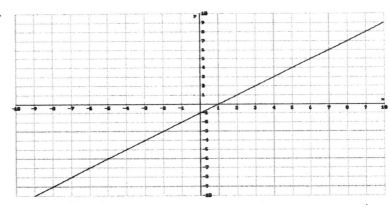

    c. x-intercept: (1, 0)
          y-intercept: (0, –1)

9. a.

x	–4	0	4
y	8	0	–8

b.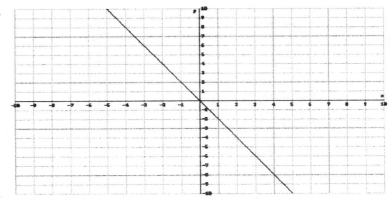

    c. x-intercept: (0, 0)
          y-intercept: (0, 0)

11. a.

x	–5	0	5
y	0	–5	–10

b.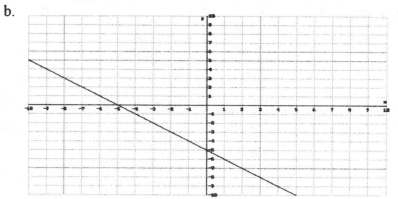

c. x-intercept: $(-5, 0)$
     y-intercept: $(0, -5)$

13. a.

x	–4	0	2
y	–5	3	7

b.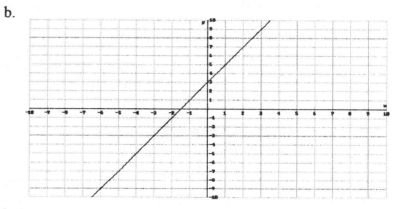

c. x-intercept: $\left(-\dfrac{3}{2}, 0\right)$
     y-intercept: $(0, 3)$

15. a.

x	–6	0	6
y	–7	–4	–1

b.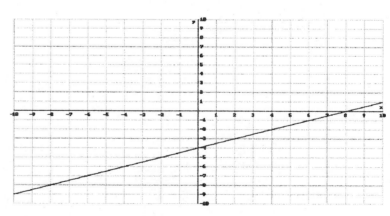

c. x-intercept: $(8, 0)$
     y-intercept: $(0, -4)$

17. a. $y = 20x$

b.

x	y
0	0
10	200
20	400
40	800

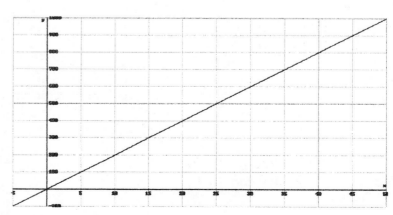

19.　a.　　$y = -800 + 5x$

b.

x	y
0	−800
50	−550
100	−300
200	200

c.
$$0 = -800 + 5x$$
$$\underline{+800 \quad +800}$$
$$800 = 5x$$
$$\frac{800}{5} = \frac{5x}{5}$$
$$160 = x$$

He must type 160 pages to break even.

21.
$$2x - 6 = -10$$
$$\underline{+6 \quad +6}$$
$$2x = -4$$
$$\frac{2x}{2} = \frac{-4}{2}$$
$$x = -2$$

23.
$$3c^2 - 7 = 5$$
$$\underline{+7 \quad +7}$$
$$3c^2 = 12$$
$$\frac{3c^2}{3} = \frac{12}{3}$$
$$c^2 = 4$$
$$c = \pm\sqrt{4}$$
$$c = \pm 2$$

25.
$$3h^2 = 15$$
$$\frac{3h^2}{3} = \frac{15}{3}$$
$$h^2 = 5$$
$$h = \pm\sqrt{5}$$

27.
$$\frac{x^2}{2} + 3 = 6$$
$$\underline{-3 \quad -3}$$
$$\frac{x^2}{2} = 3$$
$$2\left(\frac{x^2}{2}\right) = (3)2$$
$$x^2 = 6$$
$$x = \pm\sqrt{6}$$

29. $$\frac{5}{6} - \frac{4}{9} = \frac{5 \cdot 3}{6 \cdot 3} - \frac{4 \cdot 2}{9 \cdot 2} = \frac{15}{18} - \frac{8}{18} = \frac{7}{18}$$

31. $$\frac{3}{4} - \frac{3}{10} = \frac{3 \cdot 5}{4 \cdot 5} - \frac{3 \cdot 2}{10 \cdot 2} = \frac{15}{20} - \frac{6}{20} = \frac{9}{20}$$

33. $$\frac{3}{10} + \frac{7}{15} = \frac{3 \cdot 3}{10 \cdot 3} + \frac{7 \cdot 2}{15 \cdot 2} = \frac{9}{30} + \frac{14}{30} = \frac{23}{30}$$

35. $$\frac{5}{6} - \frac{5}{8} = \frac{5 \cdot 4}{6 \cdot 4} - \frac{5 \cdot 3}{8 \cdot 3} = \frac{20}{24} - \frac{15}{24} = \frac{5}{24}$$

37. $$\frac{1}{8} + \frac{7}{10} = \frac{1 \cdot 5}{8 \cdot 5} + \frac{7 \cdot 4}{10 \cdot 4} = \frac{5}{40} + \frac{28}{40} = \frac{33}{40}$$

Homework 6.5

1.   a.   Multiplied by –5, added 2
     b.   Subtract 2, divide by –5

3.   a.   Divided by 3, added 4
     b.   Subtract 4, multiply by 3

5.

$$3 - 9x = -15$$
$$\underline{-3 \qquad -3}$$
$$-9x = -18$$
$$\frac{-9x}{-9} = \frac{-18}{-9}$$
$$x = 2$$

7.

$$-3x + 7 = -26$$
$$\underline{-7 \quad -7}$$
$$-3x = -33$$
$$\frac{-3x}{-3} = \frac{-33}{-3}$$
$$x = 11$$

9.

$$\frac{-7x}{2} + 14 = 18$$
$$\underline{-14 \quad -14}$$
$$\frac{-7x}{2} = 4$$
$$2\left(\frac{-7x}{2}\right) = (4)2$$
$$-7x = 8$$
$$\frac{-7x}{-7} = \frac{8}{-7}$$
$$x = -\frac{8}{7}$$

11.

$$7 + \frac{x}{-2} = -2$$
$$\underline{-7 \qquad -7}$$
$$\frac{x}{-2} = -9$$
$$(-2)\left(\frac{x}{-2}\right) = (-9)(-2)$$
$$x = 18$$

13.

$$-17 = \frac{-6x}{5} - 5$$
$$\underline{+5 \qquad\quad +5}$$
$$-12 = \frac{-6x}{5}$$
$$5(-12) = \left(\frac{-6x}{5}\right)5$$
$$-60 = -6x$$
$$\frac{-60}{-6} = \frac{-6x}{-6}$$
$$10 = x$$
$$x = 10$$

15.

$$\frac{x+8}{-4} = -3$$
$$(-4)\left(\frac{x+8}{-4}\right) = (-3)(-4)$$
$$x + 8 = 12$$
$$\underline{-8 \quad -8}$$
$$x = 4$$

17.

$$\frac{6-5x}{8} = -3$$

$$8\left(\frac{6-5x}{8}\right) = (-3)8$$

$$6 - 5x = -24$$

$$\underline{-6 \qquad\quad -6}$$

$$-5x = -30$$

$$\frac{-5x}{-5} = \frac{-30}{-5}$$

$$x = 6$$

19. a. Time to reach −150 feet: t

b. Depth after t minutes: $-45 - 15t$
$$-150$$
$$-45 - 15t = -150$$

c.

$$-45 - 15t = -150$$

$$\underline{+45 \qquad\qquad +45}$$

$$-15t = -105$$

$$\frac{-15t}{-15} = \frac{-105}{-15}$$

$$t = 7$$

It will take 7 minutes to reach a depth of −150 feet.

21. a. Time to reach 75°: t

b. Temperature after t minutes: $350 - 11t$
$$75$$
$$350 - 11t = 75$$

c.

$$350 - 11t = 75$$

$$\underline{-350 \qquad\quad -350}$$

$$-11t = -275$$

$$\frac{-11t}{-11} = \frac{-275}{-11}$$

$$t = 25$$

It will take 25 minutes to reach room temperature.

23. a. Points for each hand: p

b. Points at 5:00: $60-12p$

$$-96$$

$$60-12p = -96$$

c.

$$60-12p = -96$$

$$\underline{-60 \qquad\quad -60}$$

$$-12p = -156$$

$$\frac{-12p}{-12} = \frac{-156}{-12}$$

$$p = 13$$

Each hand is worth 13 points.

25. a. $y = 750 - 120x$

b.

x	y
0	750
2	510
5	150
7	–90

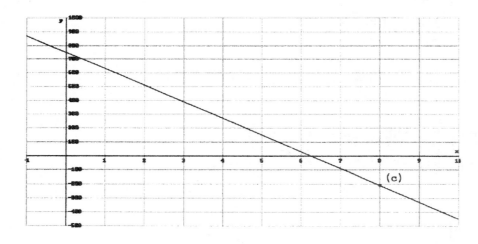

c.

$$750 - 120x = -210$$

$$\underline{-750 \qquad\qquad -750}$$

$$-120x = -960$$

$$\frac{-120x}{-120} = \frac{-960}{-120}$$

$$x = 8$$

It will take them 8 hours to reach their site.

27.

C	F
−20	−4
−10	14
0	32
10	50
20	68

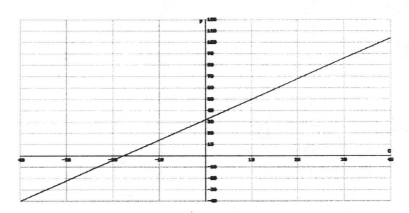

29. $-a = -(-8.2) = 8.2$

31.
$$c - d = (-3) - (-5)$$
$$= -3 + 5$$
$$= 2$$

33.
$$-3m - 2n = -3(-5) - 2(-2)$$
$$= 15 + 4$$
$$= 19$$

35.
$$\frac{3 - p}{-9 - q} = \frac{3 - (-3)}{-9 - (-9)}$$
$$= \frac{3 + 3}{-9 + 9}$$
$$= \frac{6}{0}$$
Undefined

37.
$$-1 - 2(u + 7) = -1 - 2[(-5) + 7]$$
$$= -1 - 2[2]$$
$$= -1 - 4$$
$$= -5$$

39. Multiples of 9:   9   18   27   **36**   45   54   63
Multiples of 12:   12   24   **36**   48   60   72   84
LCD = 36

41. Multiples of 10:   10   20   **30**   40   50   60   70
Multiples of 15:   15   **30**   45   60   75   90   105
LCD = 30

43. Multiples of 12:   12   24   36   48   **60**   72   84
Multiples of 15:   15   30   45   **60**   75   90   105
LCD = 60

45. Multiples of 4:   4   8   12   16   20   24   28   32   36   40
                       44   48   52   56   **60**
     Multiples of 6:   6   12   18   24   30   36   42   48   54   **60**
     Multiples of 10: 10   20   30   40   50   **60**   70
     LCD = 60

47. Multiples of 4:   4   8   12   **16**   20   24   28
     Multiples of 8:   8   **16**   24   32   40   48   56
     Multiples of 16: **16**   32   48   64   80   96   112
     LCD = 16

Chapter 6 Review

1. To convert a bar graph into a line graph, place a dot at the top of each bar and connect the dots.

3. Mean, median, mode

5. The interquartile range gives us an idea of how spread out the data values are.

7. The independent variable is plotted along the horizontal axis.

9. The solution of an equation in two variables looks like an ordered pair.

11. a. 22
    b. 1977
    c. $57 - 8 = 49$
    d. 1972, 1974, 1980, 1981, 1986, 1987

13. a. 7 8 9 10 11 13 13 14 14 15 16 17 17 18 19 21 22 22 22 22 22 24 25 25 27 28 28 29 29 30 30 30 31 32 35 35 36 37 37 41 41 46 46 47 48 55 58 59 61 142

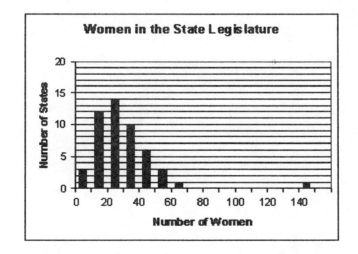

b. $\text{Mean} = \dfrac{S}{n} = \dfrac{1524}{50} = 30.48$ $\text{Median} = \dfrac{27+28}{2} = \dfrac{55}{2} = 27.5$ $\text{Mode} = 22$

c. LQ = 17   UQ = 37   IQR = 37 − 17 = 20

d.

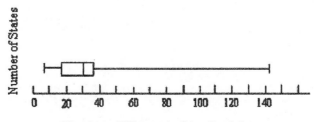

15.

t	(computation)	d
0	$16(0)^2$	0
$\frac{1}{4}$	$16\left(\frac{1}{4}\right)^2$	1
$\frac{1}{2}$	$16\left(\frac{1}{2}\right)^2$	4
$\frac{3}{4}$	$16\left(\frac{3}{4}\right)^2$	9
1	$16(1)^2$	16
$\frac{3}{2}$	$16\left(\frac{3}{2}\right)^2$	36

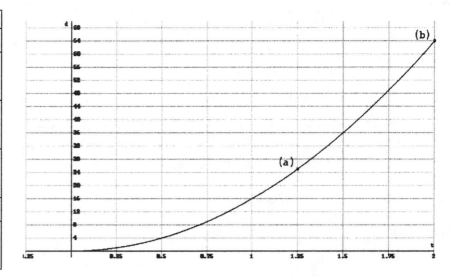

    a.   25 feet
    b.   2 seconds

17.

x	-4	-3	-2	-1	0	1	2	3	4
y	-11	-4	1	4	5	4	1	-4	-11

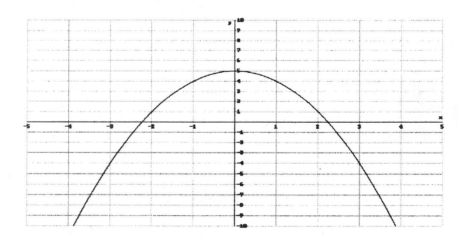

19.

x	y
-8	0
-7	1
-4	2
1	3
8	4

$$y = \sqrt{x+8}$$

21.   a.

x	-1	0	3
y	-9	-6	3

b.

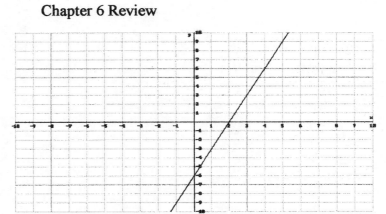

c.   x-intercept: $(2, 0)$
       y-intercept: $(0, -6)$

23.   a.

x	-3	0	6
y	-4	-2	2

b.

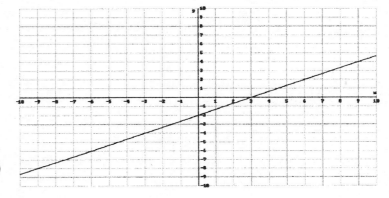

c.   x-intercept: $(3, 0)$
       y-intercept: $(0, -2)$

25.

$$5 - 3s = 8$$
$$\underline{-5 \qquad -5}$$
$$-3s = 3$$
$$\frac{-3s}{-3} = \frac{3}{-3}$$
$$s = -1$$

27.

$$6 - \frac{2a}{3} = -18$$
$$\underline{-6 \qquad -6}$$
$$-\frac{2a}{3} = -24$$
$$(-3)\left(-\frac{2a}{3}\right) = (-24)(-3)$$
$$2a = 72$$
$$\frac{2a}{2} = \frac{72}{2}$$
$$a = 36$$

29.

$$\frac{2-7w}{4} = -3$$

$$4\left(\frac{2-7w}{4}\right) = (-3)4$$

$$2-7w = -12$$

$$\underline{-2 \qquad -2}$$

$$-7w = -14$$

$$\frac{-7w}{-7} = \frac{-14}{-7}$$

$$w = 2$$

31. a. $300 - 6d$

b.

$$300 - 6d = 126$$

$$\underline{-300 \qquad -300}$$

$$-6d = -174$$

$$\frac{-6d}{-6} = \frac{-174}{-6}$$

$$d = 29$$

He will have 126 pages left after 29 days.

33.

Lisa's age	Gina's age
1	5
2	6
5	9
6	10
x	$x + 4$

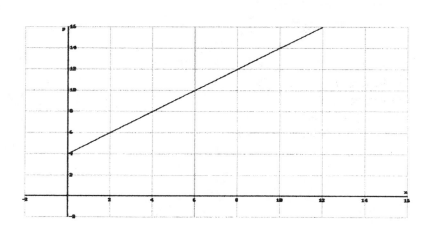

a. $x + 4$
b. (0, 4)
c. Gina was 4 years old when Lisa was born.

35. a. $y = 0.05x$

b.

x	y
250	12.5
600	30
800	40

c.

$$34 = 0.05x$$

$$\frac{34}{0.05} = \frac{0.05x}{0.05}$$

$$x = 680$$

$680,000

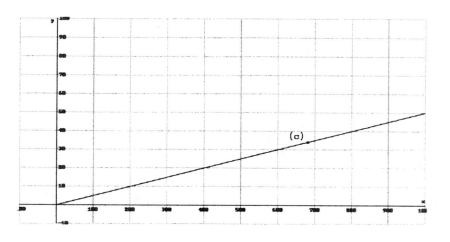

37. $\dfrac{3}{8} = \dfrac{\mathbf{6}}{16}$

39. $\dfrac{2}{3} = \dfrac{\mathbf{8}}{12}$

41. $4 + \dfrac{2}{5} = \dfrac{4 \cdot 5}{1 \cdot 5} + \dfrac{2}{5} = \dfrac{20}{5} + \dfrac{2}{5} = \dfrac{22}{5}$

43. $3 - \dfrac{3}{4} = \dfrac{3 \cdot 4}{1 \cdot 4} - \dfrac{3}{4} = \dfrac{12}{4} - \dfrac{3}{4} = \dfrac{9}{4}$

45. $\dfrac{1}{2} + \dfrac{1}{3} = \dfrac{1 \cdot 3}{2 \cdot 3} + \dfrac{1 \cdot 2}{3 \cdot 2} = \dfrac{3}{6} + \dfrac{2}{6} = \dfrac{5}{6}$

47. $\dfrac{2}{3} - \dfrac{3}{5} = \dfrac{2 \cdot 5}{3 \cdot 5} - \dfrac{3 \cdot 3}{5 \cdot 3} = \dfrac{10}{15} - \dfrac{9}{15} = \dfrac{1}{15}$

49. $\dfrac{2}{3} + \dfrac{2}{9} = \dfrac{2 \cdot 3}{3 \cdot 3} + \dfrac{2}{9} = \dfrac{6}{9} + \dfrac{2}{9} = \dfrac{8}{9}$

51. $\dfrac{8}{9} - \dfrac{5}{6} = \dfrac{8 \cdot 2}{9 \cdot 2} - \dfrac{5 \cdot 3}{6 \cdot 3} = \dfrac{16}{18} - \dfrac{15}{18} = \dfrac{1}{18}$

53.
Multiples of 4:	4	8	12	16	20	24	28	32	**36**
Multiples of 6:	6	12	18	24	30	**36**	42	48	54
Multiples of 9:	9	18	27	**36**	45	54	63	72	81

LCD = 36

Homework 7.1 A

1. The denominator of a fraction tells us into how many pieces the whole is divided. The numerator of a fraction tells us how many pieces of the whole to take.

3. A rational number is a positive or negative fraction or zero. For example, $-9, 17, \frac{4}{3}, -\frac{1}{77}$, and 0 are all rational numbers.

5. True; $\frac{4}{5} > \frac{3}{5}$

7. True; $-5 = \frac{-5}{1}$

9. False; $\frac{-3}{-7} = \frac{3}{7}$

11. $\frac{3}{9} = \frac{1}{3}$

13. $\frac{3}{5}$

15. $\frac{7}{16}$

17. $\frac{2}{8} = \frac{1}{4}$

19. $\dfrac{\text{number of infected trees}}{\text{number of trees}} = \frac{7}{18}$

21. $\dfrac{\text{number of five - year - olds}}{\text{number of kindergarteners}} = \frac{18}{18 + 9 + 2} = \frac{18}{29}$

23. $\dfrac{\text{money spent on rent}}{\text{salary}} = \frac{750}{1200} = \frac{5}{8}$

25. (fraction of a can)(number of days) $= \frac{1}{3}(14) = \frac{14}{3} = 4\frac{2}{3}$ cans

27.

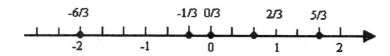

29.

31.

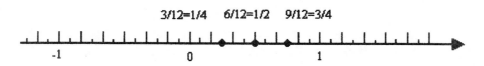

33.   Number of first-graders immunized: x

$$\text{Fraction immunized} = \frac{\text{number immunized}}{\text{number of first - graders}} = \frac{x}{84}$$

35.   Number of Digitronics employees: x

$$\text{Fraction of employees enrolled} = \frac{\text{number enrolled}}{\text{number of Digitronic s employees}} = \frac{138}{x}$$

37.   Number of gallons the tank holds: x

$$\text{Car's fuel efficiency} = \frac{\text{distance}}{\text{gallons used}} = \frac{480}{3x} = \frac{160}{x}$$

39.   Area: A Width: W

$$\text{Length} = \frac{\text{Area}}{\text{Width}} = \frac{A}{W}$$

41.

$$-3x - 8 = 7$$
$$\underline{+8 \ +8}$$
$$-3x = 15$$
$$\frac{-3x}{-3} = \frac{15}{-3}$$
$$x = -5$$

43.

$$-15 = 12 - 9w$$
$$\underline{-12 \ -12}$$
$$-27 = -9w$$
$$\frac{-27}{-9} = \frac{-9w}{-9}$$
$$3 = w$$
$$w = 3$$

45.

$$\frac{c}{8} + 6 = -2$$
$$\underline{-6 \quad -6}$$
$$\frac{c}{8} = -8$$
$$8\left(\frac{c}{8}\right) = (-8)8$$
$$c = -64$$

47.

$$14 - \frac{2v}{5} = -4$$
$$\underline{-14 \qquad -14}$$
$$-\frac{2v}{5} = -18$$
$$(-5)\left(-\frac{2v}{5}\right) = (-18)(-5)$$
$$2v = 90$$
$$\frac{2v}{2} = \frac{90}{2}$$
$$v = 45$$

49.

$$-2 + \frac{8K}{3} = -12$$

$$\underline{+2 \qquad\qquad +2}$$

$$\frac{8K}{3} = -10$$

$$3\left(\frac{8K}{3}\right) = (-10)3$$

$$8K = -30$$

$$\frac{8K}{8} = \frac{-30}{8}$$

$$K = -\frac{15}{4}$$

51. $38.2 + 21.7 \approx 38 + 22 = 60$

b

53. $8.6 - 2.9 \approx 9 - 3 = 6$

b

55. $201 + 412.9 \approx 200 + 400 = 600$

c

57. $497 - 26.7 \approx 500 - 0 = 500$

a

59.

$$209 + 142 + 48.64 \approx 200 + 150 + 50$$
$$= 400$$

a

Homework 7.1 B

1. The Fundamental Principle of Fractions is used to reduce or build fractions.

3. factor; term

5. $\dfrac{15}{12} = \dfrac{\cancel{3}\cdot 5}{\cancel{3}\cdot 4} = \dfrac{5}{4}$

7. $\dfrac{-56}{63} = \dfrac{-\cancel{7}\cdot 8}{\cancel{7}\cdot 9} = -\dfrac{8}{9}$

9. $\dfrac{48}{-36} = \dfrac{\cancel{4}\cdot\cancel{3}\cdot 4}{-\cancel{4}\cdot\cancel{3}\cdot 3} = -\dfrac{4}{3}$

11. $-\dfrac{120}{240} = -\dfrac{120\cdot 1}{120\cdot 2} = -\dfrac{1}{2}$

13. $\dfrac{15}{3x} = \dfrac{\cancel{3}\cdot 5}{\cancel{3}\cdot x} = \dfrac{5}{x}$

15. $\dfrac{6}{2w} = \dfrac{\cancel{2}\cdot 3}{\cancel{2}\cdot w} = \dfrac{3}{w}$

17. $\dfrac{24b}{14} = \dfrac{\cancel{2}\cdot 12b}{\cancel{2}\cdot 7} = \dfrac{12b}{7}$

19. $\dfrac{-5z}{6z} = \dfrac{-5\cdot\cancel{z}}{6\cdot\cancel{z}} = \dfrac{-5}{6}$

21. $\dfrac{t}{17t} = \dfrac{1 \cdot \cancel{t}}{17 \cdot \cancel{t}} = \dfrac{1}{17}$

23. $\dfrac{5u}{120uv} = \dfrac{\cancel{5} \cdot \cancel{u}}{\cancel{5} \cdot 24\cancel{u}v} = \dfrac{1}{24v}$

25. $\dfrac{16ab}{-10ab} = \dfrac{2 \cdot 8\cancel{a}\cancel{b}}{2 \cdot -5\cancel{a}\cancel{b}} = \dfrac{8}{-5} = \dfrac{-8}{5}$

27. $\dfrac{3a^2}{27a} = \dfrac{\cancel{3}\cancel{a}a}{\cancel{3} \cdot 9\cancel{a}} = \dfrac{a}{9}$

29. $\dfrac{-9y^3z}{42yz} = \dfrac{\cancel{3} \cdot -3y\cancel{y}\cancel{y}\cancel{z}}{\cancel{3} \cdot 14\cancel{y}\cancel{z}} = \dfrac{-3y^2}{14}$

31. $\dfrac{8u^3v^2}{12v^2w} = \dfrac{\cancel{4} \cdot 2uuu\cancel{v}\cancel{v}}{\cancel{4} \cdot 3\cancel{v}\cancel{v}w} = \dfrac{2u^3}{3w}$

33. a. Since we cannot cancel terms, the fraction cannot be reduced.

    b. $\dfrac{\cancel{a} \cdot 4}{\cancel{a} \cdot 5} = \dfrac{4}{5}$

35. a. $\dfrac{2 \cdot m}{4 \cdot n} = \dfrac{\cancel{2} \cdot m}{\cancel{2} \cdot 2 \cdot n} = \dfrac{m}{2n}$

    b. Since we cannot cancel terms, the fraction cannot be reduced.

37. a. Since we cannot cancel terms, the fraction cannot be reduced.

    b. $\dfrac{z(-3)}{z(+9)} = \dfrac{\cancel{z}(\cancel{3})(-1)}{\cancel{z}(\cancel{3})(3)} = \dfrac{-1}{3}$

39. a. $\dfrac{u(-v)}{u(v)} = \dfrac{\cancel{u}(\cancel{v})(-1)}{\cancel{u}(\cancel{v})} = -1$

    b. Since we cannot cancel terms, the fraction cannot be reduced.

41. a. Since we cannot cancel terms, the fraction cannot be reduced.

    b. $\dfrac{3\cancel{x}\cancel{y}}{2\cancel{x}\cancel{y}} = \dfrac{3}{2}$

43. a. $30 = 6 \cdot \mathbf{5}$
Building factor = 5

    b. $\dfrac{-5}{6} = \dfrac{-5 \cdot 5}{6 \cdot 5} = \dfrac{-25}{30}$

45. a. $9 = 1 \cdot \mathbf{9}$
Building factor = 9

    b. $3 = \dfrac{3}{1} = \dfrac{3 \cdot 9}{1 \cdot 9} = \dfrac{27}{9}$

47. a. $4z = 2 \cdot 2z = 2z \cdot \mathbf{2}$
Building factor = 2

    b. $\dfrac{1}{2z} = \dfrac{1 \cdot 2}{2z \cdot 2} = \dfrac{2}{4z}$

49. a. $11m = 11 \cdot m = m \cdot \mathbf{11}$
Building factor = 11

    b. $\dfrac{8}{m} = \dfrac{8 \cdot 11}{m \cdot 11} = \dfrac{88}{11m}$

51. a. $19b = 19 \cdot \mathbf{b}$
Building factor = b

    b. $\dfrac{-4}{19} = \dfrac{-4 \cdot b}{19 \cdot b} = \dfrac{-4b}{19b}$

53. a. $7pq = 7 \cdot q \cdot \mathbf{p}$
Building factor = p

    b. $\dfrac{p}{7q} = \dfrac{p \cdot p}{7q \cdot p} = \dfrac{p^2}{7pq}$

55. a. $2dw = w \cdot \mathbf{2d}$
Building factor = 2d

    b. $\dfrac{-9d}{w} = \dfrac{-9d \cdot 2d}{w \cdot 2d} = \dfrac{-18d^2}{2dw}$

57. a. $18x = 18 \cdot \mathbf{x}$
Building factor = x

    b. $\dfrac{x}{18} = \dfrac{x \cdot x}{18 \cdot x} = \dfrac{x^2}{18x}$

59. a. $12g = 4 \cdot \mathbf{3g}$
Building factor = 3g

    b. $\dfrac{-3g}{4} = \dfrac{-3g \cdot 3g}{4 \cdot 3g} = \dfrac{-9g^2}{12g}$

61.

a. $ab^2 = ab \cdot b$
Building factor $= b$

b. $\dfrac{-1}{ab} = \dfrac{-1 \cdot b}{ab \cdot b} = \dfrac{-b}{ab^2}$

63.

a. $2n = 1 \cdot \boldsymbol{2n}$
Building factor $= 2n$

b. $\dfrac{5n}{1} = \dfrac{5n \cdot 2n}{1 \cdot 2n} = \dfrac{10n^2}{2n}$

65.

a. $24rw^3 = 2 \cdot 12 \cdot r \cdot w^3 = 2r \cdot \boldsymbol{12w^3}$
Building factor $= 12w^3$

b. $\dfrac{-3w}{2r} = \dfrac{-3w \cdot 12w^3}{2r \cdot 12w^3} = \dfrac{-36w^4}{24rw^3}$

67.

a. 10,000 papers
b. 1980
c. 1965
d. 1955

69.

t	v
0.75	16
1	12
1.2	10
1.8	$6.\overline{6}$
2	6
2.5	4.8
3	4
4	3

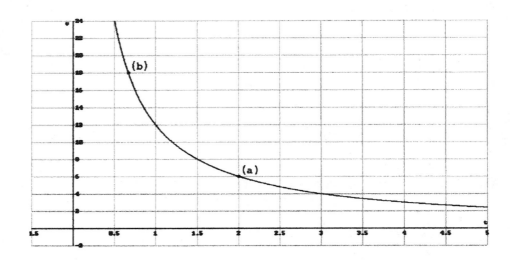

a. 6 mph

b. $\dfrac{2}{3}$ hour

71.

$-8z + 6z - (-5z) - 3z = -2z + 5z - 3z$
$\qquad\qquad\qquad\qquad\quad = 3z - 3z$
$\qquad\qquad\qquad\qquad\quad = 0$

73.

$6y - 8y - (-2y) - 3y = -2y + 2y - 3y$
$\qquad\qquad\qquad\qquad\quad = 0 - 3y$
$\qquad\qquad\qquad\qquad\quad = -3y$

75.

$6 + 2H - 9H - 6 - (-6H)$
$= 6 - 7H - 6 + 6H$
$= 6 - 6 - 7H + 6H$
$= -H$

77.

$4 - 3a + 6ab - (-8a) - 10 - 9ab + 2a$
$= 4 - 3a + 6ab + 8a - 10 - 9ab + 2a$
$= 6ab - 9ab - 3a + 8a + 2a + 4 - 10$
$= -3ab + 7a - 6$

79. $48.1 + 31.6 \approx 50 + 30 = 80$
b

81. $9.4 - 2.3 \approx 9 - 2 = 7$
b

83. $311 + 292.4 \approx 300 + 300 = 600$
b

85. $6265 - 236.6 \approx 6300 - 300 = 6000$
c

87. $28.1 + 31.4 + 1.93 \approx 28 + 30 + 2 = 60$
a

Homework 7.2 A

1. Divide

3. $\dfrac{3}{5}$ of $3\dfrac{3}{4} = \dfrac{3}{5} \cdot \dfrac{15}{4} = \dfrac{3}{\cancel{5}} \cdot \dfrac{3 \cdot \cancel{5}}{4} = \dfrac{3 \cdot 3}{4} = \dfrac{9}{4} = 2\dfrac{1}{4}$ acres

5. $\dfrac{1}{3} \cdot \dfrac{2}{9} = \dfrac{1 \cdot 2}{3 \cdot 9} = \dfrac{2}{27}$

7. $\dfrac{8}{3} \cdot \dfrac{1}{\cancel{8}} = \dfrac{1}{3}$

9. $\dfrac{-2}{d} \cdot \dfrac{c}{2} = \dfrac{-1 \cdot \cancel{2}}{d} \cdot \dfrac{c}{\cancel{2}} = \dfrac{-c}{d}$

11. $\dfrac{8}{v} \cdot \dfrac{-u}{6} = \dfrac{2 \cdot 4}{v} \cdot \dfrac{-u}{2 \cdot 3} = \dfrac{-4u}{3v}$

13. $9 \cdot \dfrac{x}{6} = \dfrac{3 \cdot 3}{1} \cdot \dfrac{x}{3 \cdot 2} = \dfrac{3x}{2}$

15. $\dfrac{w}{8} \cdot \dfrac{7}{z} = \dfrac{7w}{8z}$

17. $\dfrac{-r}{3} \cdot \dfrac{17}{r} = \dfrac{-1 \cdot \cancel{r}}{3} \cdot \dfrac{17}{\cancel{r}} = \dfrac{-17}{3}$

19. $\dfrac{-2}{a} \cdot \dfrac{3d}{4a} = \dfrac{-1 \cdot \cancel{2}}{a} \cdot \dfrac{3d}{\cancel{2} \cdot 2a} = \dfrac{-3d}{2a^2}$

21. $-5c \cdot \dfrac{3}{20} = \dfrac{5 \cdot -c}{1} \cdot \dfrac{3}{5 \cdot 4} = \dfrac{-3c}{4}$

23. $\dfrac{5}{6m^2} \cdot 2m = \dfrac{5}{2 \cdot 3mm} \cdot \dfrac{2m}{1} = \dfrac{5}{3m}$

25. $\dfrac{12s}{5r} \cdot \dfrac{2r}{3s} = \dfrac{3 \cdot 4\cancel{s}}{5\cancel{r}} \cdot \dfrac{2\cancel{r}}{3\cancel{s}} = \dfrac{8}{5}$

27. $\dfrac{-k^2}{14j} \cdot \dfrac{7j}{2k} = \dfrac{-\cancel{k}k}{7 \cdot 2\cancel{j}} \cdot \dfrac{7\cancel{j}}{2\cancel{k}} = \dfrac{-k}{4}$

29.

$\dfrac{24u^3}{7v} \cdot \dfrac{21v}{8u} = \dfrac{8 \cdot 3\cancel{u}uu}{7\cancel{v}} \cdot \dfrac{7 \cdot 3\cancel{v}}{8\cancel{u}}$

$= \dfrac{9u^2}{1}$

$= 9u^2$

31. $\dfrac{21r^2}{4rs} \cdot \dfrac{16s}{5r} = \dfrac{21\cancel{r}\cancel{r}}{4\cancel{r}\cancel{s}} \cdot \dfrac{4 \cdot 4\cancel{s}}{5\cancel{r}} = \dfrac{84}{5}$

33. $\left(\dfrac{2}{3z}\right)^2 = \dfrac{2}{3z} \cdot \dfrac{2}{3z} = \dfrac{4}{9z^2}$

35. $\left(\dfrac{-n}{7}\right)^2 = \dfrac{-n}{7} \cdot \dfrac{-n}{7} = \dfrac{n^2}{49}$

37. $\left(\dfrac{5c}{2d}\right)^2 = \dfrac{5c}{2d} \cdot \dfrac{5c}{2d} = \dfrac{25c^2}{4d^2}$

39. $\left(\dfrac{-h}{3k}\right)^3 = \dfrac{-h}{3k} \cdot \dfrac{-h}{3k} \cdot \dfrac{-h}{3k} = \dfrac{-h^3}{27k^3}$

41. a.

x	-4	0	4
y	8	4	0

b.

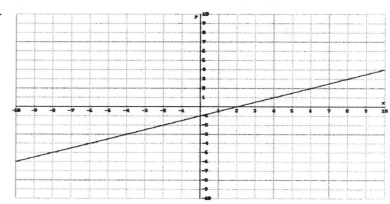

c. x-intercept: $(4, 0)$
     y-intercept: $(0, 4)$

43. a.

x	-6	0	6
y	-4	-1	2

b.

c. x-intercept: $(2, 0)$
     y-intercept: $(0, -1)$

45. $2.1(9.7) \approx 2(10) = 20$
     b

47. $31(19) \approx 30(20) = 600$
     a

49. $8.2(29) \approx 8(30) = 240$
     a

51. $0.8(618) \approx 0.8(600) = 480$
     b

53. $0.2(0.82) \approx 0.2(0.8) = 0.16$
     a

Homework 7.2 B

1. We find the reciprocal of a fraction by interchanging the numerator and denominator.

3.
$$12\frac{1}{2} \div \frac{5}{6} = \frac{25}{2} \div \frac{5}{6}$$
$$= \frac{25}{2} \cdot \frac{6}{5}$$
$$= \frac{\cancel{5} \cdot 5}{2} \cdot \frac{2 \cdot 3}{\cancel{5}}$$
$$= \frac{15}{1} = 15 \text{ lots}$$

5.
$$\frac{-16}{9} \div \frac{4}{3} = \frac{-16}{9} \cdot \frac{3}{4}$$
$$= \frac{-4 \cdot \cancel{4}}{\cancel{3} \cdot 3} \cdot \frac{\cancel{3}}{\cancel{4}}$$
$$= \frac{-4}{3}$$

7.
$$\frac{-15}{4} \div \frac{-3}{8} = \frac{-15}{4} \cdot \frac{8}{-3}$$
$$= \frac{+\cancel{3} \cdot 5}{\cancel{4}} \cdot \frac{\cancel{4} \cdot 2}{+\cancel{3}}$$
$$= 10$$

9.
$$\frac{14}{21} \div -4 = \frac{14}{21} \cdot \frac{1}{-4}$$
$$= \frac{2 \cdot \cancel{7}}{3 \cdot \cancel{7}} \cdot \frac{1}{-2 \cdot \cancel{2}}$$
$$= -\frac{1}{6}$$

11.
$$\frac{5}{3a} \div \frac{15}{a} = \frac{5}{3a} \cdot \frac{a}{15}$$
$$= \frac{\cancel{5}}{3\cancel{a}} \cdot \frac{\cancel{a}}{\cancel{5} \cdot 3}$$
$$= \frac{1}{9}$$

13.
$$\frac{24}{5h} \div \frac{-8}{5} = \frac{24}{5h} \cdot \frac{5}{-8}$$
$$= \frac{\cancel{8} \cdot 3}{\cancel{5}h} \cdot \frac{\cancel{5}}{-\cancel{8}}$$
$$= \frac{-3}{h}$$

15.
$$\frac{-9}{2p} \div -36p = \frac{-9}{2p} \cdot \frac{1}{-36p}$$
$$= \frac{+\cancel{9}}{2p} \cdot \frac{1}{+\cancel{9} \cdot 4p}$$
$$= \frac{1}{8p^2}$$

17.
$$\frac{-z^2}{2} \div \frac{z}{4} = \frac{-z^2}{2} \cdot \frac{4}{z}$$
$$= \frac{-\cancel{z}z}{\cancel{2}} \cdot \frac{\cancel{2} \cdot 2}{\cancel{z}}$$
$$= -2z$$

19.

$$\frac{24x^3}{7} \div \frac{8x^4}{21} = \frac{24x^3}{7} \cdot \frac{21}{8x^4}$$

$$= \frac{8 \cdot 3xxx}{7} \cdot \frac{7 \cdot 3}{8xxxx}$$

$$= \frac{9}{x}$$

21.

$$\frac{-15}{c^5} \div \frac{20}{9c^3} = \frac{-15}{c^5} \cdot \frac{9c^3}{20}$$

$$= \frac{5 \cdot -3}{ccccc} \cdot \frac{9ccc}{5 \cdot 4}$$

$$= \frac{-27}{4c^2}$$

23.

$$\frac{a}{b} \div \frac{c}{d} = \frac{a}{b} \cdot \frac{d}{c}$$

$$= \frac{ad}{bc}$$

25.

$$\frac{a}{c} \div \frac{b}{c} = \frac{a}{c} \cdot \frac{c}{b}$$

$$= \frac{a}{c} \cdot \frac{c}{b}$$

$$= \frac{a}{b}$$

27.

$$2A^2 \div \frac{6A}{B^2} = \frac{2A^2}{1} \cdot \frac{B^2}{6A}$$

$$= \frac{2AA}{1} \cdot \frac{BB}{2 \cdot 3A}$$

$$= \frac{AB^2}{3}$$

29.

$$\frac{3T^2}{4K^3} \div 9T^3 = \frac{3T^2}{4K^3} \cdot \frac{1}{9T^3}$$

$$= \frac{3TT}{4K^3} \cdot \frac{1}{3 \cdot 3TTT}$$

$$= \frac{1}{12K^3T}$$

31.

$$\frac{4m}{3k} \div \frac{-8}{9mk} = \frac{4m}{3k} \cdot \frac{9mk}{-8}$$

$$= \frac{4m}{3k} \cdot \frac{3 \cdot 3mk}{4 \cdot -2}$$

$$= -\frac{3m^2}{2}$$

33.

$$\frac{-2p^2r^2}{p} \div \frac{-4r^3}{p^2r} = \frac{-2p^2r^2}{p} \cdot \frac{p^2r}{-4r^3}$$

$$= \frac{+2p^2rr}{p} \cdot \frac{ppr}{+2 \cdot 2rrr}$$

$$= \frac{p^3}{2}$$

35. $w = p^2 + 1$

37. $K = -\dfrac{2}{3}m$

39. $-4a^3 + 6a^3 = 2a^3$

41.
$$-8p - 9p^2 - 3p = -9p^2 - 8p - 3p$$
$$= -9p^2 - 11p$$

43. $3b(-4b^2) = -12b^3$

45. $W^3 + W^3 = 2W^3$

47.
 a. $2m + 2m + 2m + 2m = 8m$
 b. $(2m)(2m)(2m)(2m) = 16m^4$

49. $8.8 \div 2.9 \approx 9 \div 3 = 3$
 b

51. $\dfrac{391}{21} \approx \dfrac{400}{20} = 20$
 b

53. $\dfrac{809}{3.9} \approx \dfrac{800}{4} = 200$
 c

55. $\dfrac{15,898}{19.5} \approx \dfrac{16,000}{20} = 800$
 c

57. $120 \div 0.4 = 1200 \div 4 = 300$
 c

Homework 7.3 A

1.   Like fractions are fractions that have the same denominator.

3.   To add like fractions, we add the numerators and keep the same denominator. To add like terms, we add the coefficients and keep the same variable factor.

5. $\dfrac{2}{9} + \dfrac{5}{9} = \dfrac{2+5}{9} = \dfrac{7}{9}$

7. $\dfrac{11}{12} - \dfrac{5}{12} = \dfrac{11-5}{12} = \dfrac{6}{12} = \dfrac{1}{2}$

9. $\dfrac{3}{c} + \dfrac{4}{c} = \dfrac{3+4}{c} = \dfrac{7}{c}$

11. $\dfrac{5}{3q} - \dfrac{1}{3q} = \dfrac{5-1}{3q} = \dfrac{4}{3q}$

13. $\dfrac{-7a}{6} + \dfrac{11a}{6} = \dfrac{-7a+11a}{6} = \dfrac{4a}{6} = \dfrac{2a}{3}$

15. $\dfrac{-5s}{4k} - \dfrac{3s}{4k} = \dfrac{-5s-3s}{4k} = \dfrac{-8s}{4k} = \dfrac{-2s}{k}$

17. $\dfrac{-5r}{8} + \dfrac{5r}{8} = \dfrac{-5r+5r}{8} = \dfrac{0}{8} = 0$

19. $\dfrac{6p}{5} - \dfrac{2}{5} = \dfrac{6p-2}{5}$

21. $\dfrac{10}{3v} - \dfrac{4h}{3v} = \dfrac{10-4h}{3v}$

23. $\dfrac{-4v^3}{9} - \dfrac{4v^3}{9} = \dfrac{-4v^3-4v^3}{9} = \dfrac{-8v^3}{9}$

25. $\dfrac{8}{ab} + \left(\dfrac{-9}{ab}\right) = \dfrac{8+(-9)}{ab} = \dfrac{-1}{ab}$

27.
$$\dfrac{-3}{x^2} - \left(\dfrac{-1}{x^2}\right) = \dfrac{-3-(-1)}{x^2}$$
$$= \dfrac{-3+1}{x^2}$$
$$= \dfrac{-2}{x^2}$$

29. $\dfrac{5}{b+3} + \dfrac{2}{b+3} = \dfrac{5+2}{b+3} = \dfrac{7}{b+3}$

31. $\dfrac{3c}{n-2} - \dfrac{8c}{n-2} = \dfrac{3c-8c}{n-2} = \dfrac{-5c}{n-2}$

33.
$$\dfrac{-9m}{2m-3} + \dfrac{5m}{2m-3} = \dfrac{-9m+5m}{2m-3}$$
$$= \dfrac{-4m}{2m-3}$$

35. $\dfrac{4w}{w+3} - \dfrac{3}{w+3} = \dfrac{4w-3}{w+3}$

37.

$$\frac{2x-1}{x} + \frac{2x+4}{x} = \frac{2x-1+2x+4}{x}$$

$$= \frac{4x+3}{x}$$

39.

$$\frac{u+5}{3u} - \frac{2u-1}{3u} = \frac{u+5-(2u-1)}{3u}$$

$$= \frac{u+5-2u+1}{3u}$$

$$= \frac{-u+6}{3u}$$

41. a. $\dfrac{1}{5} + \dfrac{3}{5} = \dfrac{1+3}{5} = \dfrac{4}{5}$

b. $\dfrac{1}{5} - \dfrac{3}{5} = \dfrac{1-3}{5} = \dfrac{-2}{5}$

c. $\dfrac{1}{5} \cdot \dfrac{3}{5} = \dfrac{1\cdot 3}{5\cdot 5} = \dfrac{3}{25}$

d. $\dfrac{1}{5} \div \dfrac{3}{5} = \dfrac{1}{5} \cdot \dfrac{5}{3} = \dfrac{1}{\cancel{5}} \cdot \dfrac{\cancel{5}}{3} = \dfrac{1}{3}$

43. a. $\dfrac{a}{3} + \dfrac{b}{3} = \dfrac{a+b}{3}$

b. $\dfrac{a}{3} - \dfrac{b}{3} = \dfrac{a-b}{3}$

c. $\dfrac{a}{3} \cdot \dfrac{b}{3} = \dfrac{a\cdot b}{3\cdot 3} = \dfrac{ab}{9}$

d. $\dfrac{a}{3} \div \dfrac{b}{3} = \dfrac{a}{3} \cdot \dfrac{3}{b} = \dfrac{a}{\cancel{3}} \cdot \dfrac{\cancel{3}}{b} = \dfrac{a}{b}$

45. Answers may vary.

47. a.

x	-3	0	2
y	8	5	3

b.

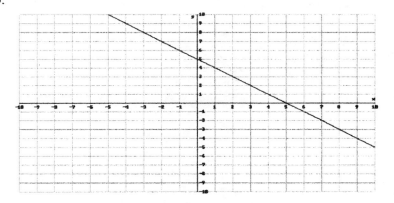

c. x-intercept: $(5, 0)$
    y-intercept: $(0, 5)$

49. a.

x	-8	0	6
y	-3	1	4

b.

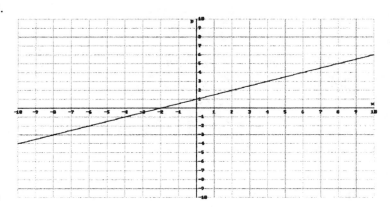

c. x-intercept: $(-2, 0)$
    y-intercept: $(0, 1)$

51. $55.6 + 28.73 = 84.33$

53. $946.2 - 627.53 = 318.67$

55. $6.4 \times 37.6 = 240.64$

57. $257.4 \div 46.8 = 5.5$

59. $621.7 \times 0.8 = 497.36$

Homework 7.3 B

1. smallest; divide into evenly

3.
| Multiples of 8: | 8 | 16 | 24 | 32 | **40** | 48 | 56 | 64 |
| Multiples of 10: | 10 | 20 | 30 | **40** | 50 | 60 | 70 | 80 |

LCD = 40

5. Yes; 16 is the smallest number than 8 and 16 divide into evenly.

7. No; 4*xy* is a factor of each denominator, not a multiple.

9. $180 = 2 \cdot 2 \cdot 3 \cdot 3 \cdot 5$

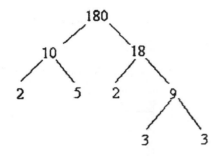

11. $360 = 2 \cdot 2 \cdot 2 \cdot 3 \cdot 3 \cdot 5$

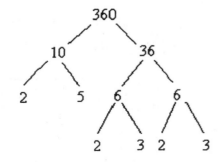

13. $15 = 3 \cdot 5$; $9 = 3 \cdot 3$
     LCD $= 3 \cdot 3 \cdot 5 = 45$

15. $14 = 2 \cdot 7$; $35 = 5 \cdot 7$
     LCD $= 2 \cdot 5 \cdot 7 = 70$

17. $10 = 2 \cdot 5; \ 6 = 2 \cdot 3; \ 12 = 2 \cdot 2 \cdot 3$
$LCD = 2 \cdot 2 \cdot 3 \cdot 5 = 60$

19. $15 = 3 \cdot 5; \ 24 = 2 \cdot 2 \cdot 2 \cdot 3; \ 35 = 5 \cdot 7$
$LCD = 2 \cdot 2 \cdot 2 \cdot 3 \cdot 5 \cdot 7 = 840$

21.
$3x = 3 \cdot x$
$x = x$
$LCD = 3 \cdot x = 3x$

23.
$2g = 2 \cdot g$
$6 = 2 \cdot 3$
$LCD = 2 \cdot 3 \cdot g = 6g$

25.
$24t = 2 \cdot 2 \cdot 2 \cdot 3 \cdot t$
$18t = 2 \cdot 3 \cdot 3 \cdot t$
$LCD = 2 \cdot 2 \cdot 2 \cdot 3 \cdot 3 \cdot t = 72t$

27.
$4n = 2 \cdot 2 \cdot n$
$2n^2 = 2 \cdot n \cdot n$
$LCD = 2 \cdot 2 \cdot n \cdot n = 4n^2$

29.
$5a = 5 \cdot a$
$3b = 3 \cdot b$
$LCD = 3 \cdot 5 \cdot a \cdot b = 15ab$

31.
$w^3 = w \cdot w \cdot w$
$w = w$
$LCD = w \cdot w \cdot w = w^3$

33.
$3xy = 3 \cdot x \cdot y$
$4y^2 = 2 \cdot 2 \cdot y \cdot y$
$LCD = 2 \cdot 2 \cdot 3 \cdot x \cdot y \cdot y = 12xy^2$

35.
$8uv^2 = 2 \cdot 2 \cdot 2 \cdot u \cdot v \cdot v$
$12u^2v = 2 \cdot 2 \cdot 3 \cdot u \cdot u \cdot v$
$LCD = 2 \cdot 2 \cdot 2 \cdot 3 \cdot u \cdot u \cdot v \cdot v = 24u^2v^2$

37.
$5p = 5 \cdot p$
$25p^2 = 5 \cdot 5 \cdot p \cdot p$
$125p^3 = 5 \cdot 5 \cdot 5 \cdot p \cdot p \cdot p$
$LCD = 5 \cdot 5 \cdot 5 \cdot p \cdot p \cdot p = 125p^3$

39.
$qr = q \cdot r$
$rs = r \cdot s$
$sq = s \cdot q$
$LCD = q \cdot r \cdot s = qrs$

41. a. $\dfrac{3}{5a} + \dfrac{b}{5a} = \dfrac{3+b}{5a}$

    b. $\dfrac{3}{5a} \cdot \dfrac{b}{5a} = \dfrac{3 \cdot b}{5a \cdot 5a} = \dfrac{3b}{25a^2}$

43. a.

$$\dfrac{3r}{4p^2} \div \dfrac{9r}{2p} = \dfrac{3r}{4p^2} \cdot \dfrac{2p}{9r}$$

$$= \dfrac{3\cancel{r}}{2 \cdot 2\cancel{p} \cdot p} \cdot \dfrac{2\cancel{p}}{3 \cdot 3\cancel{r}}$$

$$= \dfrac{1}{6p}$$

    b. $\dfrac{3r}{4p^2} \cdot \dfrac{9r}{2p} = \dfrac{3r \cdot 9r}{4p^2 \cdot 2p} = \dfrac{27r^2}{8p^3}$

45. $\sqrt{36+64} = \sqrt{100} = 10$

47. $5\sqrt{81} = 5(9) = 45$

49.

$$7 - 3\sqrt{169} = 7 - 3(13)$$
$$= 7 - 39$$
$$= -32$$

51. $\sqrt{\dfrac{225}{16}} = \dfrac{15}{4}$

53. $\dfrac{8+\sqrt{16}}{4} = \dfrac{8+4}{4} = \dfrac{12}{4} = 3$

55. $\sqrt{4} < \sqrt{6} < \sqrt{9}$

a

57. $\sqrt{20} > \sqrt{16}$

c

59. $\sqrt{64} < \sqrt{73} < \sqrt{81}$

a

61. $\sqrt{81} < \sqrt{98} < \sqrt{100}$

b

63. $\sqrt{17.8} > \sqrt{16}$

c

Homework 7.4

1.   We cannot add or subtract pieces that are not the same size.

3.   a. $LCD = 2 \cdot 3 = 6$

      b. $\dfrac{1}{2} = \dfrac{1 \cdot 3}{2 \cdot 3} = \dfrac{3}{6} ; \dfrac{2}{3} = \dfrac{2 \cdot 2}{3 \cdot 2} = \dfrac{4}{6}$

      c. $\dfrac{3}{6} + \dfrac{4}{6} = \dfrac{3+4}{6} = \dfrac{7}{6}$

5.   a. $4 = 2 \cdot 2 ; \; 3 = 3$

           $LCD = 2 \cdot 2 \cdot 3 = 12$

      b. $\dfrac{3}{4} = \dfrac{3 \cdot 3}{4 \cdot 3} = \dfrac{9}{12} ; \dfrac{2}{3} = \dfrac{2 \cdot 4}{3 \cdot 4} = \dfrac{8}{12}$

      c. $\dfrac{9}{12} - \dfrac{8}{12} = \dfrac{9-8}{12} = \dfrac{1}{12}$

7.   a. $6 = 2 \cdot 3 ; \; 9 = 3 \cdot 3$

           $LCD = 2 \cdot 3 \cdot 3 = 18$

      b. $\dfrac{5}{6} = \dfrac{5 \cdot 3}{6 \cdot 3} = \dfrac{15}{18} ; \dfrac{2}{9} = \dfrac{2 \cdot 2}{9 \cdot 2} = \dfrac{4}{18}$

      c. $\dfrac{15}{18} + \dfrac{4}{18} = \dfrac{15+4}{18} = \dfrac{19}{18}$

9.   a. $5 = 5 ; \; 10 = 2 \cdot 5$

           $LCD = 2 \cdot 5 = 10$

      b. $\dfrac{3}{5} = \dfrac{3 \cdot 2}{5 \cdot 2} = \dfrac{6}{10}$

      c. $\dfrac{6}{10} - \dfrac{7}{10} = \dfrac{6-7}{10} = \dfrac{-1}{10}$

11.   a. $3 = 3 ; \; 5 = 5$

           $LCD = 3 \cdot 5 = 15$

      b. $\dfrac{a}{3} = \dfrac{a \cdot 5}{3 \cdot 5} = \dfrac{5a}{15} ; \dfrac{b}{5} = \dfrac{b \cdot 3}{5 \cdot 3} = \dfrac{3b}{15}$

      c. $\dfrac{5a}{15} + \dfrac{3b}{15} = \dfrac{5a+3b}{15}$

13.   a. $4 = 2 \cdot 2 ; \; 10 = 2 \cdot 5$

           $LCD = 2 \cdot 2 \cdot 5 = 20$

      b. $\dfrac{m}{4} = \dfrac{m \cdot 5}{4 \cdot 5} = \dfrac{5m}{20}$

           $\dfrac{9}{10} = \dfrac{9 \cdot 2}{10 \cdot 2} = \dfrac{18}{20}$

      c. $\dfrac{5m}{20} - \dfrac{18}{20} = \dfrac{5m-18}{20}$

15.   a. $8 = 2 \cdot 2 \cdot 2 ; \; 4 = 2 \cdot 2$

           $LCD = 2 \cdot 2 \cdot 2 = 8$

      b. $\dfrac{5z}{4} = \dfrac{5z \cdot 2}{4 \cdot 2} = \dfrac{10z}{8}$

      c. $\dfrac{-3w}{8} - \dfrac{10z}{8} = \dfrac{-3w-10z}{8}$

17.   a. $9 = 3 \cdot 3 ; \; 6 = 2 \cdot 3$

           $LCD = 2 \cdot 3 \cdot 3 = 18$

      b. $\dfrac{-2}{9} = \dfrac{-2 \cdot 2}{9 \cdot 2} = \dfrac{-4}{18}$

           $\dfrac{5v}{6} = \dfrac{5v \cdot 3}{6 \cdot 3} = \dfrac{15v}{18}$

      c. $\dfrac{-4}{18} + \dfrac{15v}{18} = \dfrac{-4+15v}{18}$

19.   a. $4 = 2 \cdot 2 ; \; 5 = 5$

           $LCD = 2 \cdot 2 \cdot 5 = 20$

      b. $\dfrac{b^2}{4} = \dfrac{b^2 \cdot 5}{4 \cdot 5} = \dfrac{5b^2}{20}$

           $\dfrac{-b}{5} = \dfrac{-b \cdot 4}{5 \cdot 4} = \dfrac{-4b}{20}$

      c. $\dfrac{5b^2}{20} + \dfrac{-4b}{20} = \dfrac{5b^2-4b}{20}$

21.   a. $LCD = 9$

      b. $1 = \dfrac{1 \cdot 9}{1 \cdot 9} = \dfrac{9}{9}$

      c. $\dfrac{9}{9} + \dfrac{y}{9} = \dfrac{9+y}{9}$

23. a. $LCD = 4$

   b. $c = \dfrac{c \cdot 4}{1 \cdot 4} = \dfrac{4c}{4}$

   c. $\dfrac{4c}{4} - \dfrac{c}{4} = \dfrac{4c - c}{4} = \dfrac{3c}{4}$

25. a. $10 = 2 \cdot 5\,;\ 8 = 2 \cdot 2 \cdot 2$
     $LCD = 2 \cdot 2 \cdot 2 \cdot 5 = 40$

   b. $\dfrac{x}{10} = \dfrac{x \cdot 4}{10 \cdot 4} = \dfrac{4x}{40}$

     $\dfrac{x}{8} = \dfrac{x \cdot 5}{8 \cdot 5} = \dfrac{5x}{40}$

   c. $\dfrac{4x}{40} - \dfrac{5x}{40} = \dfrac{4x - 5x}{40} = \dfrac{-x}{40}$

27. a. $LCD = 3p$

   b. $\dfrac{2}{p} = \dfrac{2 \cdot 3}{p \cdot 3} = \dfrac{6}{3p}$

     $\dfrac{q}{3} = \dfrac{q \cdot p}{3 \cdot p} = \dfrac{pq}{3p}$

   c. $\dfrac{6}{3p} + \dfrac{pq}{3p} = \dfrac{6 + pq}{3p}$

29. a. $LCD = ab$

   b. $\dfrac{1}{a} = \dfrac{1 \cdot b}{a \cdot b} = \dfrac{b}{ab}\,;\ \dfrac{2}{b} = \dfrac{2 \cdot a}{b \cdot a} = \dfrac{2a}{ab}$

   c. $\dfrac{b}{ab} - \dfrac{2a}{ab} = \dfrac{b - 2a}{ab}$

31. a. $xy = x \cdot y\,;\ y = y$
     $LCD = x \cdot y = xy$

   b. $\dfrac{3}{y} = \dfrac{3 \cdot x}{y \cdot x} = \dfrac{3x}{xy}$

   c. $\dfrac{5}{xy} - \dfrac{3x}{xy} = \dfrac{5 - 3x}{xy}$

33. a. $st = s \cdot t\,;\ t = t$
     $LCD = s \cdot t = st$

   b. $\dfrac{s}{t} = \dfrac{s \cdot s}{t \cdot s} = \dfrac{s^2}{st}$

   c. $\dfrac{1}{st} + \dfrac{s^2}{st} = \dfrac{1 + s^2}{st}$

35. a. $LCD = ab$

   b. $\dfrac{-3}{a} = \dfrac{-3 \cdot b}{a \cdot b} = \dfrac{-3b}{ab}$

     $\dfrac{2a}{b} = \dfrac{2a \cdot a}{b \cdot a} = \dfrac{2a^2}{ab}$

   c. $\dfrac{-3b}{ab} - \dfrac{2a^2}{ab} = \dfrac{-3b - 2a^2}{ab}$

37. a. $LCD = v$

   b. $2 = \dfrac{2 \cdot v}{1 \cdot v} = \dfrac{2v}{v}$

   c. $\dfrac{2v}{v} + \dfrac{1}{v} = \dfrac{2v + 1}{v}$

39. a. $z = z\,;\ 2z = 2 \cdot z$
     $LCD = 2 \cdot z = 2z$

   b. $\dfrac{2}{z} = \dfrac{2 \cdot 2}{z \cdot 2} = \dfrac{4}{2z}$

   c. $\dfrac{4}{2z} - \dfrac{1}{2z} = \dfrac{4 - 1}{2z} = \dfrac{3}{2z}$

41. a. $5n = 5 \cdot n\,;\ 4n = 2 \cdot 2 \cdot n$
     $LCD = 2 \cdot 2 \cdot 5 \cdot n = 20n$

   b. $\dfrac{-2}{5n} = \dfrac{-2 \cdot 4}{5n \cdot 4} = \dfrac{-8}{20n}$

     $\dfrac{q}{4n} = \dfrac{q \cdot 5}{4n \cdot 5} = \dfrac{5q}{20n}$

   c. $\dfrac{-8}{20n} + \dfrac{5q}{20n} = \dfrac{-8 + 5q}{20n}$

43.　a.　$s = s \, ; \ s^2 = s \cdot s$
　　　$\text{LCD} = s \cdot s = s^2$

　　b.　$\dfrac{-6}{s} = \dfrac{-6 \cdot s}{s \cdot s} = \dfrac{-6s}{s^2}$

　　c.　$\dfrac{-6s}{s^2} + \dfrac{3}{s^2} = \dfrac{-6s + 3}{s^2}$

45.　a.　$4x^2 = 2 \cdot 2 \cdot x \cdot x$
　　　$6xz = 2 \cdot 3 \cdot x \cdot z$
　　　$\text{LCD} = 2 \cdot 2 \cdot 3 \cdot x \cdot x \cdot z = 12x^2 z$

　　b.　$\dfrac{z}{4x^2} = \dfrac{z \cdot 3z}{4x^2 \cdot 3z} = \dfrac{3z^2}{12x^2 z}$

　　　$\dfrac{1}{6xz} = \dfrac{1 \cdot 2x}{6xz \cdot 2x} = \dfrac{2x}{12x^2 z}$

　　c.　$\dfrac{3z^2}{12x^2 z} - \dfrac{2x}{12x^2 z} = \dfrac{3z^2 - 2x}{12x^2 z}$

47.

　　a.　$\dfrac{1}{x} + \dfrac{x}{2} = \dfrac{1 \cdot 2}{x \cdot 2} + \dfrac{x \cdot x}{2 \cdot x}$

　　　　　$= \dfrac{2}{2x} + \dfrac{x^2}{2x}$

　　　　　$= \dfrac{2 + x^2}{2x}$

　　b.　$\dfrac{1}{x} \div \dfrac{x}{2} = \dfrac{1}{x} \cdot \dfrac{2}{x}$

　　　　　$= \dfrac{1 \cdot 2}{x \cdot x}$

　　　　　$= \dfrac{2}{x^2}$

　　c.　$\dfrac{1}{x} \cdot \dfrac{x}{2} = \dfrac{1 \cdot \cancel{x}}{\cancel{x} \cdot 2} = \dfrac{1}{2}$

　　d.

　　　$\dfrac{1}{x} - \dfrac{x}{2} = \dfrac{1 \cdot 2}{x \cdot 2} - \dfrac{x \cdot x}{2 \cdot x}$

　　　　　$= \dfrac{2}{2x} - \dfrac{x^2}{2x}$

　　　　　$= \dfrac{2 - x^2}{2x}$

49.　a.　$\dfrac{3}{2a} - \dfrac{2}{a^2} = \dfrac{3 \cdot a}{2a \cdot a} - \dfrac{2 \cdot 2}{a^2 \cdot 2}$

　　　　　$= \dfrac{3a}{2a^2} - \dfrac{4}{2a^2}$

　　　　　$= \dfrac{3a - 4}{2a^2}$

　　b.　$\dfrac{3}{2a} \cdot \dfrac{2}{a^2} = \dfrac{3}{\cancel{2}a} \cdot \dfrac{\cancel{2}}{a^2} = \dfrac{3}{a^3}$

　　c.　$\dfrac{3}{2a} + \dfrac{2}{a^2} = \dfrac{3 \cdot a}{2a \cdot a} + \dfrac{2 \cdot 2}{a^2 \cdot 2}$

　　　　　$= \dfrac{3a}{2a^2} + \dfrac{4}{2a^2}$

　　　　　$= \dfrac{3a + 4}{2a^2}$

　　d.　$\dfrac{3}{2a} \div \dfrac{2}{a^2} = \dfrac{3}{2a} \cdot \dfrac{a^2}{2}$

　　　　　$= \dfrac{3}{2\cancel{a}} \cdot \dfrac{\cancel{a} \cdot a}{2}$

　　　　　$= \dfrac{3a}{4}$

51. $\frac{1}{4} = 0.25 = 25\%$

b

53. $\frac{1}{3} = 0.\overline{3} = 33.\overline{3}\% \approx 33\%$

c

55. $\frac{2}{3} = 0.\overline{6} = 66.\overline{6}\% \approx 66\%$

b

57. $\frac{31}{101} \approx \frac{30}{100} = 30\%$

a

59. $\frac{205}{793} \approx \frac{1}{4} = 25\%$

b

Homework 7.5

1. its reciprocal

3.
$$\frac{1}{5}x = 10$$
$$5\left(\frac{1}{5}x\right) = (10)5$$
$$x = 50$$

5.
$$\frac{2}{3}y = 12$$
$$\frac{3}{2}\left(\frac{2}{3}y\right) = (12)\frac{3}{2}$$
$$1 \cdot y = \frac{12}{1} \cdot \frac{3}{2}$$
$$y = \frac{36}{2} = 18$$

7.
$$\frac{5}{4}a = -25$$
$$\frac{4}{5}\left(\frac{5}{4}a\right) = (-25)\frac{4}{5}$$
$$1 \cdot a = \frac{-25}{1} \cdot \frac{4}{5}$$
$$a = \frac{-5 \cdot \cancel{5}}{1} \cdot \frac{4}{\cancel{5}} = -20$$

9.
$$\frac{3}{2}p = -7$$
$$\frac{2}{3}\left(\frac{3}{2}p\right) = (-7)\frac{2}{3}$$
$$1 \cdot p = \frac{-7}{1} \cdot \frac{2}{3}$$
$$p = \frac{-14}{3}$$

11.
$$\frac{-2}{9}k = \frac{3}{5}$$
$$\frac{9}{-2}\left(\frac{-2}{9}k\right) = \left(\frac{3}{5}\right)\frac{9}{-2}$$
$$1 \cdot k = \frac{3}{5} \cdot \frac{9}{-2}$$
$$k = \frac{-27}{10}$$

13.
$$\frac{-4}{3}u = \frac{-8}{21}$$
$$\frac{3}{-4}\left(\frac{-4}{3}u\right) = \left(\frac{-8}{21}\right)\frac{3}{-4}$$
$$1 \cdot u = \frac{+4 \cdot 2}{\cancel{3} \cdot 7} \cdot \frac{\cancel{3}}{+\cancel{4}}$$
$$u = \frac{2}{7}$$

15.
$$\frac{8}{5}q = \frac{2}{7}$$
$$\frac{35}{1}\left(\frac{8}{5}q\right) = \left(\frac{2}{7}\right)\frac{35}{1}$$
$$56q = 10$$
$$q = \frac{10}{56} = \frac{5}{28}$$

17.

$$\frac{-5m}{3} = \frac{7}{2}$$

$$\frac{6}{1}\left(\frac{-5}{3}m\right) = \left(\frac{7}{2}\right)\frac{6}{1}$$

$$-10m = 21$$

$$m = -\frac{21}{10}$$

19.

$$\frac{2}{b} = \frac{-3}{4}$$

$$\frac{4b}{1}\left(\frac{2}{b}\right) = \left(\frac{-3}{4}\right)\frac{4b}{1}$$

$$8 = -3b$$

$$\frac{-8}{3} = b$$

$$b = \frac{-8}{3}$$

21.

$$\frac{1}{5} = \frac{3}{10x}$$

$$\frac{10x}{1}\left(\frac{1}{5}\right) = \left(\frac{3}{10x}\right)\frac{10x}{1}$$

$$2x = 3$$

$$x = \frac{3}{2}$$

23.

$$\frac{-1}{8x} = \frac{5}{6}$$

$$\frac{24x}{1}\left(\frac{-1}{8x}\right) = \left(\frac{5}{6}\right)\frac{24x}{1}$$

$$-3 = 20x$$

$$-\frac{3}{20} = x$$

$$x = -\frac{3}{20}$$

25.

$$\frac{n-9}{3} = -2$$

$$\frac{3}{1}\left(\frac{n-9}{3}\right) = (-2)\frac{3}{1}$$

$$n-9 = -6$$

$$\underline{+9 \quad +9}$$

$$n = 3$$

27.

$$1 = \frac{11}{n+1}$$

$$\frac{n+1}{1}(1) = \left(\frac{11}{n+1}\right)\frac{n+1}{1}$$

$$n+1 = 11$$

$$\underline{-1 \quad -1}$$

$$n = 10$$

29. Darcy's weekly pay: x

$$\frac{2}{9}x = 130$$

$$\frac{9}{2}\left(\frac{2}{9}x\right) = (130)\frac{9}{2}$$

$$1x = \frac{2 \cdot 65}{1} \cdot \frac{9}{2}$$

$$x = 585$$

Darcy makes $585 each week.

31. Number of votes cast: x

$$\frac{2}{3}x = 48$$

$$\frac{3}{2}\left(\frac{2}{3}x\right) = (48)\frac{3}{2}$$

$$1x = \frac{2 \cdot 24}{1} \cdot \frac{3}{2}$$

$$x = 72$$

There were 72 votes cast.

33. Number of fish in lake: x

$$\frac{3}{20}x = 24$$

$$\frac{20}{3}\left(\frac{3}{20}x\right) = (24)\frac{20}{3}$$

$$1x = \frac{\cancel{3}\cdot 8}{1}\cdot\frac{20}{\cancel{3}}$$

$$x = 160$$

There are 160 fish in the lake.

35. a.

$$\frac{2b}{c} - \frac{1}{2c} = \frac{2b\cdot 2}{c\cdot 2} - \frac{1}{2c}$$

$$= \frac{4b}{2c} - \frac{1}{2c}$$

$$= \frac{4b-1}{2c}$$

b.

$$\frac{2b}{c} \div \frac{1}{2c} = \frac{2b}{c}\cdot\frac{2c}{1}$$

$$= \frac{2b}{\cancel{c}}\cdot\frac{2\cancel{c}}{1}$$

$$= 4b$$

37. a.

$$\frac{v}{p}\cdot\frac{2q}{v} = \frac{\cancel{v}}{p}\cdot\frac{2q}{\cancel{v}}$$

$$= \frac{2q}{p}$$

b.

$$\frac{v}{p} + \frac{2q}{v} = \frac{v\cdot v}{p\cdot v} + \frac{2q\cdot p}{v\cdot p}$$

$$= \frac{v^2}{pv} + \frac{2pq}{pv}$$

$$= \frac{v^2 + 2pq}{pv}$$

39. a.

$$\frac{a}{9b} \div \frac{a^2}{6b} = \frac{a}{9b}\cdot\frac{6b}{a^2}$$

$$= \frac{\cancel{a}}{3\cdot\cancel{3}\cdot\cancel{b}}\cdot\frac{2\cdot\cancel{3}\cdot\cancel{b}}{\cancel{a}\cdot a}$$

$$= \frac{2}{3a}$$

b.

$$\frac{a}{9b} + \frac{a^2}{6b} = \frac{a\cdot 2}{9b\cdot 2} + \frac{a^2\cdot 3}{6b\cdot 3}$$

$$= \frac{2a}{18b} + \frac{3a^2}{18b}$$

$$= \frac{2a + 3a^2}{18b}$$

41.

$$m(n+p) = (-4)[(-3)+6]$$

$$= (-4)[3]$$

$$= -12$$

43.

$$m - n - q = (-4) - (-3) - (-2)$$

$$= -4 + 3 + 2$$

$$= -1 + 2$$

$$= 1$$

45.

$$m - nq = (-4) - (-3)(-2)$$

$$= (-4) - 6$$

$$= -10$$

47.

$$\frac{p}{q} - m = \frac{6}{-2} - (-4)$$

$$= -3 + 4$$

$$= 1$$

49.

$$\frac{p-q}{p+m} = \frac{6-(-2)}{6+(-4)}$$

$$= \frac{6+2}{2}$$

$$= \frac{8}{2}$$

$$= 4$$

51. 27% of $28 \approx \frac{1}{4}(28) = 7$

53. 34.5% of $96 \approx \frac{1}{3}(96) = 32$

55. 72% of $120 \approx \frac{3}{4}(120) = 90$

57. 68% of $4500 \approx \frac{2}{3}(4500) = 3000$

59. 52% of $320 \approx \frac{1}{2}(320) = 160$

Chapter 7 Review

1. False; $\dfrac{4}{6} = \dfrac{2}{3}$

3. False; We need a common denominator to add or subtract fractions, but not to multiply or divide fractions.

5. $-\dfrac{2}{3}$ is a negative number, but $\dfrac{-2}{-3} = \dfrac{2}{3}$ is a positive number.

7. The opposite of 2 is –2, and the reciprocal of 2 is $\dfrac{1}{2}$.

9. $\dfrac{a}{b} + \dfrac{c}{d} = \dfrac{ad}{bd} + \dfrac{bc}{bd} = \dfrac{ad + bc}{bd}$

11. Number of cats at HappyCat Boarding: x

    Fraction with special diets $= \dfrac{\text{number with special diets}}{\text{number of cats boarded}} = \dfrac{5}{x}$

13. Number of students in class: x

    Number of bags used = (number of students)(bags per student) $= x \cdot \dfrac{1}{3} = \dfrac{x}{3}$

15.

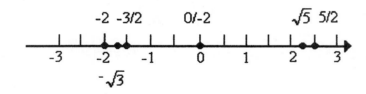

17. a.
    $$\begin{aligned} \frac{5 - 2^3}{5 - 2} &= \frac{5 - 8}{3} \\ &= \frac{-3}{3} \\ &= -1 \end{aligned}$$

    b.
    $$\begin{aligned} \frac{5(-2)^3}{5(-2)} &= \frac{5(-8)}{-10} \\ &= \frac{-40}{-10} \\ &= 4 \end{aligned}$$

19. a. $\sqrt{\dfrac{100}{25}} = \sqrt{4} = 2$

    b. $\dfrac{\sqrt{100}}{\sqrt{25}} = \dfrac{10}{5} = 2$

21. $\dfrac{4x}{8} = \dfrac{\cancel{4} \cdot x}{\cancel{4} \cdot 2} = \dfrac{x}{2}$

a and c

23. $\dfrac{-2bc}{bc} = \dfrac{-2 \cdot \cancel{b} \cdot \cancel{c}}{\cancel{b} \cdot \cancel{c}} = -2$

b and d

25. $\dfrac{2}{3} \cdot \dfrac{-5}{4} = \dfrac{\cancel{2}}{3} \cdot \dfrac{-5}{\cancel{2} \cdot 2} = \dfrac{-5}{6}$

27.
$$\dfrac{4}{5} \div \dfrac{2}{3} = \dfrac{4}{5} \cdot \dfrac{3}{2}$$
$$= \dfrac{\cancel{2} \cdot 2}{5} \cdot \dfrac{3}{\cancel{2}}$$
$$= \dfrac{6}{5}$$

29. $\dfrac{3}{7} + \dfrac{4}{7} = \dfrac{3+4}{7} = \dfrac{7}{7} = 1$

31. $\dfrac{7}{9} - \dfrac{4}{9} = \dfrac{7-4}{9} = \dfrac{3}{9} = \dfrac{1}{3}$

33. $\dfrac{3x}{4} \cdot \dfrac{-6}{5x^2} = \dfrac{3\cancel{x}}{\cancel{2} \cdot 2} \cdot \dfrac{-3 \cdot \cancel{2}}{5\cancel{x} \cdot x} = \dfrac{-9}{10x}$

35.
$$\dfrac{-6m^2}{10n} \div \dfrac{n}{3m} = \dfrac{-6m^2}{10n} \cdot \dfrac{3m}{n}$$
$$= \dfrac{\cancel{2} \cdot -3m^2}{\cancel{2} \cdot 5n} \cdot \dfrac{3m}{n}$$
$$= \dfrac{-9m^3}{5n^2}$$

37. $\dfrac{3}{2r} + \dfrac{-1}{2r} = \dfrac{3 + -1}{2r} = \dfrac{\cancel{2}}{\cancel{2}r} = \dfrac{1}{r}$

39.
$$\dfrac{4v}{7w^2} - \dfrac{-3v}{7w^2} = \dfrac{4v - (-3v)}{7w^2}$$
$$= \dfrac{4v + 3v}{7w^2}$$
$$= \dfrac{\cancel{7}v}{\cancel{7}w^2}$$
$$= \dfrac{v}{w^2}$$

41.

$$\frac{1}{3} + \frac{1}{6} = \frac{1 \cdot 2}{3 \cdot 2} + \frac{1}{6}$$

$$= \frac{2}{6} + \frac{1}{6}$$

$$= \frac{2+1}{6}$$

$$= \frac{3}{6}$$

$$= \frac{\cancel{3}}{\cancel{3} \cdot 2}$$

$$= \frac{1}{2}$$

43.

$$\frac{1}{2} - \frac{1}{2a} = \frac{1 \cdot a}{2 \cdot a} - \frac{1}{2a}$$

$$= \frac{a}{2a} - \frac{1}{2a}$$

$$= \frac{a-1}{2a}$$

45.

$$2 - \frac{3}{2x} = \frac{2 \cdot 2x}{1 \cdot 2x} - \frac{3}{2x}$$

$$= \frac{4x}{2x} - \frac{3}{2x}$$

$$= \frac{4x-3}{2x}$$

47.

$$\frac{6}{ab} + \frac{4}{b^2} = \frac{6 \cdot b}{ab \cdot b} + \frac{4 \cdot a}{b^2 \cdot a}$$

$$= \frac{6b}{ab^2} + \frac{4a}{ab^2}$$

$$= \frac{6b + 4a}{ab^2}$$

49.

$$-\left(\frac{21b}{7}\right)^4 = -\left(\frac{3 \cdot \cancel{7}b}{\cancel{7}}\right)^4$$

$$= -(3b)^4$$

$$= -(3b)(3b)(3b)(3b)$$

$$= -81b^4$$

51.

$$\frac{1}{7}w = 2$$

$$7\left(\frac{1}{7}w\right) = (2)7$$

$$w = 14$$

53.

$$\frac{5}{2}a = \frac{-4}{3}$$

$$\frac{2}{5}\left(\frac{5}{2}a\right) = \left(\frac{-4}{3}\right)\frac{2}{5}$$

$$1a = \frac{-4}{3} \cdot \frac{2}{5}$$

$$a = \frac{-8}{15}$$

55.

$$\frac{9}{2d} = -5$$

$$2d\left(\frac{9}{2d}\right) = (-5)2d$$

$$9 = -10d$$

$$\frac{9}{-10} = \frac{-10d}{-10}$$

$$d = \frac{-9}{10}$$

57.

$$\frac{8}{3} = \frac{-16}{c}$$

$$3c\left(\frac{8}{3}\right) = \left(\frac{-16}{c}\right)3c$$

$$8c = -48$$

$$\frac{8c}{8} = \frac{-48}{8}$$

$$c = -6$$

59.

$$1 = \frac{-5}{n-3}$$

$$(n-3)(1) = \left(\frac{-5}{n-3}\right)(n-3)$$

$$n - 3 = -5$$

$$\underline{+3 \quad +3}$$

$$n = -2$$

61.

$$5 = \frac{x+1}{6}$$

$$6(5) = \left(\frac{x+1}{6}\right)6$$

$$30 = x + 1$$

$$\underline{-1 \quad -1}$$

$$29 = x$$

$$x = 29$$

63. Number of batches: x

$$\frac{2}{3}x = 4$$

$$\frac{3}{2}\left(\frac{2}{3}x\right) = (4)\frac{3}{2}$$

$$1x = \frac{2 \cdot 2}{1} \cdot \frac{3}{2}$$

$$x = 6$$

Six batches can be made.

65. Original price of TV: x

$$\frac{3}{4}x - 150 = 825$$

$$\underline{+150 \quad +150}$$

$$\frac{3}{4}x = 975$$

$$\frac{4}{3}\left(\frac{3}{4}x\right) = (975)\frac{4}{3}$$

$$1x = \frac{3 \cdot 325}{1} \cdot \frac{4}{3}$$

$$x = 1300$$

The original price of the TV was $1300.

67. Number manufactured: x

$$\frac{8}{9}x = 1000$$

$$\frac{9}{8}\left(\frac{8}{9}x\right) = (1000)\frac{9}{8}$$

$$1x = \frac{8 \cdot 125}{1} \cdot \frac{9}{8}$$

$$x = 1125$$

1125 monitors must be manufactured.

69. $1.253 + 8.461 \approx 1.5 + 8.5 = 10$
    a

71. $\frac{629}{87} \approx \frac{630}{90} = 7$
    a

73. $82.36 - 28.41 \approx 80 - 30 = 50$
    c

75. $11.4(28.6) \approx 10(30) = 300$
    c

77. $\dfrac{3}{4} = 0.75 = 75\%$

    c

79. $0.49(813) \approx 0.50(800) = 400$

    b

81. Since $\sqrt{4} < \sqrt{7.1} < \sqrt{9},\ 2 < \sqrt{7.1} < 3$

    a

Homework 8.1 A

1.

$$P = rW$$
$$= 0.04(560)$$
$$= \$22.40$$

His salary will increase \$22.40 each week.

3.

$$P = rW$$
$$30 = r(36)$$
$$\frac{30}{36} = \frac{r(36)}{36}$$
$$0.8\overline{3} = r$$
$$r = 83.\overline{3}\%$$

He got $83.\overline{3}\%$ of the questions right.

5.

$$P = rW$$
$$21 = 0.35W$$
$$\frac{21}{0.35} = \frac{0.35W}{0.35}$$
$$60 = W$$
$$W = 60$$

There are 60 positions.

7.
a. 2%
b. $29 + 27 = 56\%$
c. $100 - 6 = 94\%$
d.

Under 20	$0.06 \times 50,000 =$	3,000
21 to 25	$0.29 \times 50,000 =$	14,500
26 to 30	$0.27 \times 50,000 =$	13,500
31 to 35	$0.17 \times 50,000 =$	8,500
36 to 40	$0.11 \times 50,000 =$	5,500
41 to 50	$0.08 \times 50,000 =$	4,000
Over 50	$0.02 \times 50,000 =$	1,000
	Total =	50,000

9.
a. 12.3%
b. Europe
c. Asia
d.

Europe	$0.712 \times 71,154 \approx$	50,662
Latin America	$0.123 \times 71,154 \approx$	8752
Asia	$0.059 \times 71,154 \approx$	4198
Oceania	$0.031 \times 71,154 \approx$	2206
Others	$0.075 \times 71,154 \approx$	5337
	Total =	71,155

11. $\dfrac{34t^2}{17t} = \dfrac{2 \cdot 17tt}{17t} = 2t$

13. $\dfrac{66bc^3}{-22bc^2} = \dfrac{2 \cdot 3 \cdot 11\cancel{b}\cancel{c}\cancel{c}c}{-2 \cdot 11\cancel{b}\cancel{c}\cancel{c}} = -3c$

15. $\dfrac{42kl^2}{14kl^2} = \dfrac{2 \cdot 3 \cdot 7 \cdot k \cdot l \cdot l}{2 \cdot 7 \cdot k \cdot l \cdot l} = 3$

17. $\dfrac{72w^2z}{42wz^2} = \dfrac{6 \cdot 12 \cdot w \cdot w \cdot z}{6 \cdot 7 \cdot w \cdot z \cdot z} = \dfrac{12w}{7z}$

19. $\dfrac{-15kt^2}{20m^2t} = \dfrac{-3 \cdot 5 \cdot k \cdot t \cdot t}{4 \cdot 5 \cdot m \cdot m \cdot t} = \dfrac{-3kt}{4m^2}$

21. Total $\approx 3 + 2 + 6 = \$11$
No

23. Total $\approx 7 + 2(6) = 7 + 12 = \$19$
Yes

25. Total $\approx 5 + 2(30) = 5 + 60 = \$65$
Yes

27.
$$\begin{aligned} \text{Total} &\approx 2(25) + 2(5) + 2(6) \\ &= 50 + 10 + 12 = \$72 \end{aligned}$$
Yes

29.
$$\begin{aligned} \text{Total} &\approx 100(3) + 100(8) \\ &= 300 + 800 = \$1100 \end{aligned}$$
No

Homework 8.1 B

1.
$$\begin{aligned} \text{Amount spent on next election} &= (1 - 0.28) \times (\text{Amount spent on last election}) \\ &= 0.72(826{,}000) \\ &= 594{,}720 \end{aligned}$$
He can spent $594,720 next time.

3.   a.
$$\begin{aligned} \text{New salary} &= (1 - 0.12) \times (\text{Old salary}) \\ &= 0.88(25{,}000) \\ &= 22{,}000 \end{aligned}$$
His new salary is $22,000

   b.
$$\begin{aligned} \text{New salary} &= (1 + 0.12) \times (\text{Old salary}) \\ &= 1.12(22{,}000) \\ &= 24{,}640 \end{aligned}$$
After 18 months, his salary was $24,640.

5.   a. $\dfrac{1200}{7500} \times 100 = 16\%$

   b. $\dfrac{864}{4800} \times 100 = 18\%$

   c.   Edith

7. Percent increase in Sterling population $= \dfrac{1200}{5000} \times 100 = 24\%$

Percent increase in school enrollment $= \dfrac{140}{460} \times 100 \approx 30\%$

The school experienced the greater percent increase in population.

9. Percent decrease in mathematics degrees $= \dfrac{667}{15,904} \times 100 \approx 4.2\%$

Percent decrease in physical sciences degrees $= \dfrac{602}{17,806} \times 100 \approx 3.4\%$

Mathematics experienced the greater percent decline in degrees awarded.

11. a.

$$\text{Sale price} = (1 - 0.15) \times (\text{List price})$$
$$= 0.85(1800)$$
$$= 1530$$
$$\text{Price after discount} = (1 - 0.10) \times (\text{Sale price})$$
$$= 0.90(1530)$$
$$= 1377$$

She will pay $1377 if she buys it through the store.

b.

$$\text{Catalog price} = (1 - 0.25) \times (\text{List price})$$
$$= 0.75(1800)$$
$$= 1350$$

She will pay $1350 if she buys it through the catalog.

c. The catalog price is the better deal.

13. a.

$$\text{Copies sold in 1991} = (1 + 0.20) \times (\text{Copies sold in 1990})$$
$$= 1.20(8000)$$
$$= 9600$$
$$\text{Copies sold in 1992} = (1 + 0.35) \times (\text{Copies sold in 1991})$$
$$= 1.35(9600)$$
$$= 12,960$$

In 1991, 9600 copies were sold. In 1992, 12,960 copies were sold.

b.

$$\text{Copies sold in 1992} = (1 + 0.55) \times (\text{Copies sold in 1990})$$
$$= 1.55(8000)$$
$$= 12,400$$

In 1992, 12,400 copies would have been sold.

c. No

15.   a.   Fall

       b. $\dfrac{-0.61}{56.62} \times 100 \approx -1.08\%$

       c. $\dfrac{-0.81}{33.62} \times 100 \approx -2.41\%$

       d.   Coca Cola (The Coca Cola stock was worth more at the beginning of the week, and both stocks lost the same amount of value.)

17.   a.

Year	Population at Start of Year	Increase in Population	Population at End of Year
1990	20,000	(0.05)(20,000) = 1000	20,000+1000 = 21,000
1991	21,000	1050	22,050
1992	22,050	1103	23,153
1993	23,153	1158	24,311
1994	24,311	1216	25,527
1995	25,527	1276	26,803
1996	26,803	1340	28,143
1997	28,143	1407	29,550
1998	29,550	1478	31,028
1999	31,028	1551	32,579
2000	32,579	1629	34,208

       b.

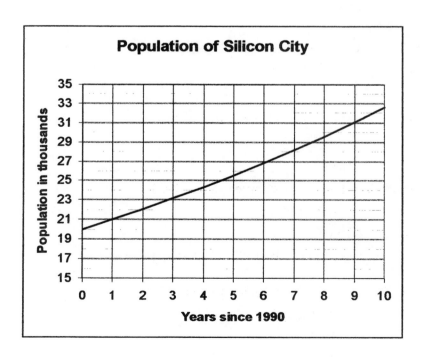

19. $$\frac{7uv}{9} = \frac{7uv \cdot uv}{9 \cdot uv} = \frac{7u^2v^2}{9uv}$$

21. $$\frac{c}{6} = \frac{c \cdot 6d^2}{6 \cdot 6d^2} = \frac{6cd^2}{36d^2}$$

23. $$1 = \frac{1 \cdot 70uv}{1 \cdot 70uv} = \frac{70uv}{70uv}$$

25. $$\frac{-6}{5} = \frac{-6 \cdot 3p^2q}{5 \cdot 3p^2q} = \frac{-18p^2q}{15p^2q}$$

27. $$\frac{9x}{11y} = \frac{9x \cdot 2xy}{11y \cdot 2xy} = \frac{18x^2y}{22xy^2}$$

29.

$$1 \text{ meter} = 39.37 \text{ inches}$$
$$10 \text{ meters} = 10(39.37) = \underline{393.7} \text{ inches}$$

31.

$$1 \text{ sandwich} = \$4.49$$
$$10 \text{ sandwiches} = 10(\$4.49) = \$44.90$$

33.

$$1 \text{ quart} = 0.946 \text{ liters}$$
$$100 \text{ quarts} = 100(0.946) = \underline{94.6} \text{ liters}$$

35.

$$1 \text{ kilogram} = 1000 \text{ grams}$$
$$0.454 \text{ kilograms} = 1000(0.454)$$
$$= \underline{454} \text{ grams}$$

37.

$$1 \text{ kilometer} = 1000 \text{ meters}$$
$$1.6093 \text{ kilometers} = 1000(1.6093)$$
$$= \underline{1609.3} \text{ meters}$$

Homework 8.2

1. a. $\dfrac{3}{4}$

   b. $\dfrac{4}{3}$

   c. $\dfrac{3}{7}$; $\dfrac{4}{7}$

   d. $\dfrac{3 \text{ sailboats}}{4 \text{ motorboats}} = \dfrac{3 \cdot 12}{4 \cdot 12} = \dfrac{36 \text{ sailboats}}{48 \text{ motorboats}}$

   48 motorboats

   e. $\dfrac{3 \text{ sailboats}}{4 \text{ motorboats}} = \dfrac{3 \cdot 15}{4 \cdot 15} = \dfrac{45 \text{ sailboats}}{60 \text{ motorboats}}$

   45 sailboats

3. $\dfrac{5 \cdot 2}{8 \cdot 2} = \dfrac{10}{16}$; $\dfrac{5 \cdot 3}{8 \cdot 3} = \dfrac{15}{24}$; $\dfrac{5 \cdot 4}{8 \cdot 4} = \dfrac{20}{32}$

   10 and 16, 15 and 24, 20 and 32

5. It is greater than 1.

7. $\dfrac{\text{pages of text}}{\text{pages of exercises}} = \dfrac{248}{186} = \dfrac{4 \cdot 62}{3 \cdot 62} = \dfrac{4}{3}$

9. a. $\dfrac{\text{calories from fat}}{\text{calories from protein}} = \dfrac{840}{240} = \dfrac{120 \cdot 7}{120 \cdot 2} = \dfrac{7}{2}$

   b. $\dfrac{\text{calories from protein}}{\text{calories from carbs}} = \dfrac{240}{960} = \dfrac{240 \cdot 1}{240 \cdot 4} = \dfrac{1}{4}$

11. $\dfrac{\text{cranberry juice}}{\text{red wine}} = \dfrac{48 \text{ oz}}{32 \text{ oz}} = \dfrac{16 \cdot 4}{16 \cdot 3} = \dfrac{3}{2}$

13. a. $\dfrac{\text{defective bulbs}}{\text{total bulbs}} = \dfrac{54}{4500} = \dfrac{18 \cdot 3}{18 \cdot 250} = \dfrac{3}{250}$

    b. $\dfrac{\text{good bulbs}}{\text{defective bulbs}} = \dfrac{250 - 3}{3} = \dfrac{247}{3}$

    c. $\dfrac{54}{4500} \times 100 = 1.2\%$

15. $\dfrac{5.6}{21.4} \approx 0.262$

17. $\dfrac{9}{14.4} = 0.625$

19. $\dfrac{11.2}{7} = 1.6$

21. $\dfrac{24.9}{41.5} = 0.6$

23. $\dfrac{432 \text{ miles}}{12 \text{ gallons}} = 36 \text{ miles per gallon}$

25. $\dfrac{40 \text{ minutes}}{5 \text{ miles}} = 8 \text{ minutes per mile}$

$\dfrac{5 \text{ miles}}{40 \text{ minutes}} = \dfrac{1}{8} \text{ mile per minute}$

27. $\dfrac{12 \cdot 15 \text{ sq feet}}{2.5 \text{ hours}} = \dfrac{180 \text{ sq feet}}{2.5 \text{ hours}} = 72 \text{ square feet per hour}$

29. a. $\dfrac{2982 - 2800 \text{ cents}}{28 \text{ dollars}} = \dfrac{182 \text{ cents}}{28 \text{ dollars}} = 6.5 \text{ cents per dollar}$

b. $\dfrac{\$0.065}{\$1} \times 100 = 6.5\%$

31. $\dfrac{10b}{5c} \cdot \dfrac{3c^2}{6bc} = \dfrac{2 \cdot \cancel{5}\cancel{b}}{\cancel{5}\cancel{c}} \cdot \dfrac{\cancel{3}\cancel{c}\cancel{c}}{2 \cdot \cancel{3}\cancel{b}\cancel{c}} = 1$

33. $\dfrac{2}{15a} \cdot \dfrac{4a}{3b} \cdot \dfrac{9b^2}{8} = \dfrac{2}{\cancel{3} \cdot 5\cancel{a}} \cdot \dfrac{4\cancel{a}}{\cancel{3}\cancel{b}} \cdot \dfrac{\cancel{3} \cdot \cancel{3}b\cancel{b}}{2 \cdot \cancel{4}} = \dfrac{b}{5}$

35. $\dfrac{-p^3}{16q^4} \cdot \dfrac{28q}{3p} \cdot \dfrac{-6q^2}{7p} = \dfrac{-p p p}{2 \cdot 2 \cdot \cancel{4} q q q q} \cdot \dfrac{4 \cdot 7 q}{3 p} \cdot \dfrac{-2 \cdot 3 q q}{7 p} = \dfrac{p}{2q}$

37. $\left(\dfrac{-2r}{s}\right)^3 = \left(\dfrac{-2r}{s}\right)\left(\dfrac{-2r}{s}\right)\left(\dfrac{-2r}{s}\right) = \dfrac{-8r^3}{s^3}$

39. $\dfrac{237}{10} = 23.7 \text{ pounds}$

41. $\dfrac{12,480}{10} = 1248 \text{ inches}$

43. $\dfrac{490}{100} = 4.9 \text{ meters}$

45. $\dfrac{946}{1000} = 0.946 \text{ liters}$

47. $\dfrac{938}{100} = 9.38 \text{ seconds}$

Homework 8.3 A

1. If two variables are proportional, their ratio is constant.

3. $\dfrac{k}{h} = \dfrac{9}{3} = 3$; $\dfrac{k}{h} = \dfrac{18}{6} = 3$

   $\dfrac{k}{h} = \dfrac{24}{8} = 3$; $\dfrac{k}{h} = \dfrac{36}{12} = 3$

   Yes; All the ratios are equal.

5. $\dfrac{z}{w} = \dfrac{0}{6} = 0$; $\dfrac{z}{w} = \dfrac{2}{10} = \dfrac{1}{5}$

   No; Since the ratios are not equal, the variables are not proportional.

7. $\dfrac{V}{Y} = \dfrac{4}{2} = 2$; $\dfrac{V}{Y} = \dfrac{9}{3} = 3$

   No; Since the ratios are not equal, the variables are not proportional.

9. $\dfrac{t}{H} = \dfrac{0.5}{2} = \dfrac{1}{4}$; $\dfrac{t}{H} = \dfrac{0.25}{4} = \dfrac{1}{16}$

   No; Since the ratios are not equal, the variables are not proportional.

11. No; The line does not go through the origin.

13. Yes; The graph is a straight line through the origin.

15. No; f is not a constant multiple of d.

17. Yes; S is a constant multiple of m.

19. Yes; amount of syrup $= \dfrac{3}{32}$ (amount of water)

21. No; thickness of book $= \dfrac{1}{25}$ (number of pages) $+ 6$

23. a. $k = \dfrac{d}{t} = \dfrac{45}{3} = 15$

    $d = 15t$

    b.

d	t
3	45
5	75
10	150

    c.

    $d = 15t$

    $300 = 15t$

    $\dfrac{300}{15} = \dfrac{15t}{15}$

    $20 = t$

    It will take them 20 hours to go 300 miles.

25.   a. $k = \dfrac{C}{w} = \dfrac{12.80}{16} = 0.8$

           $C = 0.8w$

     b.

w	C
10	8
15	12
20	16

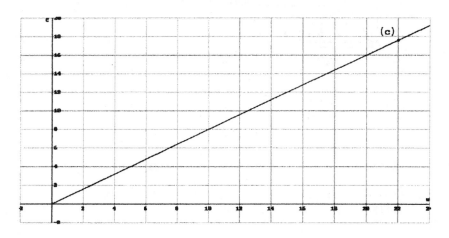

     c.

         $C = 0.8w$

         $C = 0.8(22)$

         $C = 17.60$

         A 22-pound turkey costs $17.60.

27.   Trish's commission: C; Amount of sale: S

     a. $k = \dfrac{C}{S} = \dfrac{3200}{80,000} = 0.04$

           $C = 0.04S$

     b.

         $C = 0.04S$

            $= 0.04(200,000)$

            $= 8000$

         Her commission would be $8,000.

29.   Perimeter of square: P; Length of side: s

     a. $P = 4s$

     b.

s	2	5	8	11
P	8	20	32	44

     c.

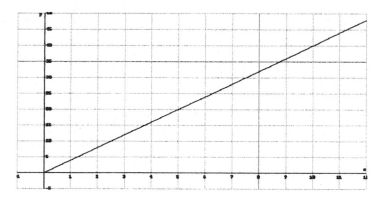

     d.   Yes; The graph is a straight line through the origin.

31.   Area of square: A; Length of side: s

a. $A = s^2$

b.

s	2	5	6	8
A	4	25	36	64

c.
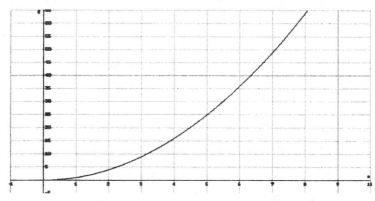

d.   No; The graph is not a straight line.

33.
$$\frac{-s^2w}{30s} \div \frac{15sw^3}{18} = \frac{-s^2w}{30s} \cdot \frac{18}{15sw^3}$$
$$= \frac{-\cancel{s}\cancel{s}w}{5 \cdot 6\cancel{s}} \cdot \frac{\cancel{3} \cdot \cancel{6}}{\cancel{3} \cdot 5\cancel{s}www}$$
$$= \frac{-1}{25w^2}$$

35.
$$\frac{8a^3}{12h^3} \div \frac{56a^2h^4}{3a^4h^2} = \frac{8a^3}{12h^3} \cdot \frac{3a^4h^2}{56a^2h^4}$$
$$= \frac{\cancel{8}\cancel{a}\cancel{a}a}{\cancel{3} \cdot 4\cancel{h}\cancel{h}h} \cdot \frac{\cancel{3}aaaa\cancel{h}\cancel{h}}{7 \cdot \cancel{8}\cancel{a}\cancel{a}hhhh}$$
$$= \frac{a^5}{28h^5}$$

37.
$$\frac{a}{b} \div \frac{c}{d} \div \frac{e}{f} = \frac{a}{b} \cdot \frac{d}{c} \div \frac{e}{f}$$
$$= \frac{ad}{bc} \cdot \frac{f}{e}$$
$$= \frac{adf}{bce}$$

39.
$$\frac{a}{b} \div \left(\frac{c}{d} \div \frac{e}{f}\right) = \frac{a}{b} \div \left(\frac{c}{d} \cdot \frac{f}{e}\right)$$
$$= \frac{a}{b} \div \left(\frac{cf}{de}\right)$$
$$= \frac{a}{b} \cdot \left(\frac{de}{cf}\right) = \frac{ade}{bcf}$$

41. $0.5(18) = \dfrac{1}{2}(18) = \dfrac{18}{2} = 9$

43. $0.125(24) = \dfrac{1}{8}(24) = \dfrac{24}{8} = 3$

45. $0.6(25) = \dfrac{3}{5}(25) = 3\left(\dfrac{25}{5}\right) = 3(5) = 15$

47. $0.75(12) = \dfrac{3}{4}(12) = 3\left(\dfrac{12}{4}\right) = 3(3) = 9$

49. $1.25(16) = \dfrac{5}{4}(16) = 5\left(\dfrac{16}{4}\right) = 5(4) = 20$

Homework 8.3 B

1.   A proportion is an equation that states two ratios are equal.

3.   Yes

5.   No; There is no equal sign.

7.   No; The left side is not a ratio.

9.   Yes

11.
$$85 \cdot 55 \overset{?}{=} 51 \cdot 97$$
$$4675 \neq 4947$$
No

13.
$$10.4 \cdot 19.5 \overset{?}{=} 6.5 \cdot 31.2$$
$$202.8 = 202.8$$
Yes

15.
$$\frac{12}{21} = \frac{28}{n}$$
$$12n = 21 \cdot 28$$
$$12n = 588$$
$$\frac{12n}{12} = \frac{588}{12}$$
$$n = 49$$

17.
$$\frac{b}{7} = \frac{9}{5}$$
$$b \cdot 5 = 7 \cdot 9$$
$$5b = 63$$
$$\frac{5b}{5} = \frac{63}{5}$$
$$b = 12.6$$

19.
$$\frac{3.6}{8} = \frac{y}{10}$$
$$3.6 \cdot 10 = 8y$$
$$36 = 8y$$
$$\frac{36}{8} = \frac{8y}{8}$$
$$4.5 = y$$
$$y = 4.5$$

21.

$$\frac{\frac{7}{3}}{3} = \frac{z}{\frac{9}{2}}$$

$$\frac{7}{3} \cdot \frac{9}{2} = 3z$$

$$\frac{21}{2} = 3z$$

$$\frac{1}{3}\left(\frac{21}{2}\right) = (3z)\frac{1}{3}$$

$$\frac{21}{6} = z$$

$$z = \frac{7}{2}$$

23. Width of poster: W

$$\frac{\text{width}}{\text{length}} : \frac{W \text{ feet}}{4 \text{ feet}} = \frac{4 \text{ inches}}{6 \text{ inches}}$$

$$6W = 4 \cdot 4$$

$$6W = 16$$

$$\frac{6W}{6} = \frac{16}{6}$$

$$W = \frac{8}{3} = 2\frac{2}{3}$$

The poster will be 2 feet and 8 inches wide.

25. Actual length: L

$$\frac{\text{actual}}{\text{map}} : \frac{L \text{ miles}}{3.25 \text{ inches}} = \frac{1.5 \text{ miles}}{0.75 \text{ inches}}$$

$$0.75L = 3.25 \cdot 1.5$$

$$0.75L = 4.875$$

$$\frac{0.75L}{0.75} = \frac{4.875}{0.75}$$

$$L = 6.5$$

The lake is actually 6.5 miles long.

27. Height of tree: h

$$\frac{\text{height}}{\text{shadow length}} : \frac{h \text{ feet}}{38 \text{ feet}} = \frac{6 \text{ feet}}{10 \text{ feet}}$$

$$10h = 38 \cdot 6$$

$$10h = 228$$

$$\frac{10h}{10} = \frac{228}{10}$$

$$h = 22.8$$

The tree is 22.8 feet tall.

29. Number of deer: n

$$\frac{\text{number tagged}}{\text{number of deer}} : \frac{9}{60} = \frac{50}{n}$$

$$9n = 60 \cdot 50$$

$$9n = 3000$$

$$\frac{9n}{9} = \frac{3000}{9}$$

$$n \approx 333$$

There are approximately 333 deer in the woods.

31.

$$\text{Area} = l \times w$$
$$= (14)(8)$$
$$= 112 \text{ square meters}$$

$$\text{Perimeter} = 2l + 2w$$
$$= 2(14) + 2(8)$$
$$= 28 + 16$$
$$= 44 \text{ meters}$$

33.   First, find the missing side.

$$a^2 + b^2 = c^2$$
$$5^2 + b^2 = 9^2$$
$$25 + b^2 = 81$$
$$\underline{-25 \qquad -25}$$
$$b^2 = 56$$
$$b = \sqrt{56} \approx 7.483$$

$$\text{Area} = \frac{1}{2}bh$$
$$= \frac{1}{2}(7.483)(5)$$
$$= 18.71 \text{ square yards}$$

$$\text{Perimeter} = a + b + c$$
$$= 5 + 7.48 + 9$$
$$= 21.48 \text{ yards}$$

35.

$$\text{Area} = \frac{1}{2}\pi r^2$$
$$= \frac{1}{2}\pi(6)^2$$
$$= 18\pi$$
$$\approx 56.55 \text{ square feet}$$

$$\text{Perimeter} = \frac{1}{2}(2\pi r) + 6 + 6$$
$$= \pi(6) + 12$$
$$\approx 30.85 \text{ feet}$$

37.   First, find the missing side.

$$a^2 + b^2 = c^2$$
$$3^2 + 4^2 = c^2$$
$$9 + 16 = c^2$$
$$25 = c^2$$
$$c = \sqrt{25} = 5$$

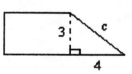

$$\text{Area} = \frac{1}{2}h(b_1 + b_2)$$
$$= \frac{1}{2}(3)(5 + 9)$$
$$= \frac{1}{2}(3)(14)$$
$$= 21 \text{ square cm}$$

$$\text{Perimeter} = 9 + 3 + 5 + 5$$
$$= 22 \text{ cm}$$

39.   First, find the missing sides.

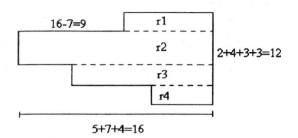

$$P = 4 + 9 + 2 + 7 + 12 + 4 + 3 + 7 + 3 + 5$$
$$= 56 \text{ feet}$$

$$A = (\text{area of } r1) + (\text{area of } r2) + (\text{area of } r3) + (\text{area of } r4)$$
$$= (2)(7) + (4)(16) + (3)(11) + (3)(4)$$
$$= 14 + 64 + 33 + 12$$
$$= 123 \text{ square feet}$$

41.
10% of $40 is $4.
5% of $40 is $4 ÷ 2 = $2.
15% of $40 is $4 + $2 = $6.

43.
10% of $32 is $3.20.
5% of $32 is $3.20 ÷ 2 = $1.60.
15% of $32 is $3.20 + $1.60 = $4.80.

45. $19.93 ≈ $20.00
10% of $20 is $2.
5% of $20 is $2 ÷ 2 = $1.
15% of $20 is $2 + $1 = $3.
b

47. $59.96 ≈ $60.00
10% of $60 is $6.
5% of $60 is $6 ÷ 2 = $3.
15% of $60 is $6 + $3 = $9.
b

49. $8.99 ≈ $9.00
10% of $9 is $0.90.
5% of $9 is $0.90 ÷ 2 = $0.45.
15% of $9 is $0.90 + $0.45 = $1.35.
b

Homework 8.4

1. b and e

3. a and c

$$\text{scale factor} = \frac{2}{4} = \frac{1}{2}$$

$$\text{scale factor} = \frac{3/2}{3} = \frac{1}{2}$$

5. a and e

$$\text{scale factor} = \frac{2}{4} = \frac{1}{2}$$

$$\text{scale factor} = \frac{3}{6} = \frac{1}{2}$$

7.

$k = 1.50$

Area will increase by a factor of

$k^2 = (1.50)^2 = \textbf{2.25}$.

enlarged area $= k^2$ (original area)

enlarged area $= (1.50)^2 (3 \cdot 5)$

enlarged area $= \textbf{33.75 square inches}$

9.

$k = 0.10$

model sail area $= k^2$ (real sail area)

model sail area $= (0.10)^2 (75)$

model sail area $= 0.75$ square feet

11.

$$\frac{\text{old dimension}}{\text{new dimension}} = \frac{1}{500}$$

$$\frac{\text{old volume}}{\text{new volume}} = \left(\frac{1}{500}\right)^3 = \frac{1}{125,000,000}$$

$1 \text{ to } 125,000,000$

13.

$$\frac{\text{area}}{\text{price}} = \frac{3\pi(8)^2}{15} \approx 40.21 \text{ square inches per dollar}$$

$$\frac{\text{area}}{\text{price}} = \frac{\pi(16)^2}{15} \approx 53.61 \text{ square inches per dollar}$$

The 16-inch pizza is the better deal because you get more pizza per dollar.

15.

$$k = \frac{60}{6} = 10$$

giant's weight $= k^3$ (human weight)

giant's weight $= 10^3 (180)$

giant's weight $= 180,000$ pounds

17.

$$-2w(3w - 2) = -2(-3)[3(-3) - 2]$$
$$= -2(-3)[-9 - 2]$$
$$= -2(-3)[-11]$$
$$= 6(-11)$$
$$= -66$$

19.
$$3x - 5x^2 - x^3 = 3(-2) - 5(-2)^2 - (-2)^3$$
$$= 3(-2) - 5(4) - (-8)$$
$$= -6 - 20 + 8$$
$$= -26 + 8$$
$$= -18$$

21.
$$\frac{1 - m^2}{1 + m^2} = \frac{1 - (-1.5)^2}{1 + (-1.5)^2}$$
$$= \frac{1 - (2.25)}{1 + (2.25)}$$
$$= \frac{-1.25}{3.25}$$
$$\approx -0.385$$

23.
$$2h^2\sqrt{6h} = 2(12.8)^2\sqrt{6(12.8)}$$
$$= 2(163.84)\sqrt{76.8}$$
$$= 327.68\sqrt{76.8}$$
$$\approx 2871.644$$

25.
$$\frac{-b + \sqrt{b^2 + 8a}}{2a} = \frac{-(-3) + \sqrt{(-3)^2 + 8(2)}}{2(2)}$$
$$= \frac{-(-3) + \sqrt{9 + 16}}{2(2)}$$
$$= \frac{-(-3) + \sqrt{25}}{2(2)}$$
$$= \frac{-(-3) + 5}{2(2)}$$
$$= \frac{3 + 5}{4}$$
$$= \frac{8}{4} = 2$$

27. 52% of $78 \approx \frac{1}{2}(80) = 40$

a

29. 32% of $291 \approx \frac{1}{3}(300) = 100$

b

31. 27% of $7.6 \approx \frac{1}{4}(8) = 2$

b

33. 65% of $302 \approx \frac{2}{3}(300) = 200$

b

35. 76.1% of $39 \approx \frac{3}{4}(40) = 30$

b

Challenge Problems:

37.
$$\frac{\text{new area}}{\text{old area}} = 4 = k^2$$
$$\frac{\text{new dimension}}{\text{old dimension}} = 2 = k$$
You should scale the dimensions by 2.

39.
$$\frac{\text{new volume}}{\text{old volume}} = 8 = k^3$$
$$\frac{\text{new dimension}}{\text{old dimension}} = 2 = k$$
You should scale the dimensions by 2.

Homework 8.5

1. a. $50

    b.   (2, 300), (10, 700)

    $$m = \frac{\Delta A}{\Delta w} = \frac{700 - 300}{10 - 2} = \frac{400}{8} = 50 \text{ dollars per week}$$

    c.   dollars/week; The slope tells us the rate at which Lynette's savings grow.

3. a. $15 - 10 = 5$ pounds; $25 - 15 = 10$ pounds

    b.   Use (4, 15) and (8, 25).

    $$m = \frac{\Delta W}{\Delta t} = \frac{25 - 15}{8 - 4} = \frac{10}{4} = 2.5 \text{ pounds per month}$$

    c.

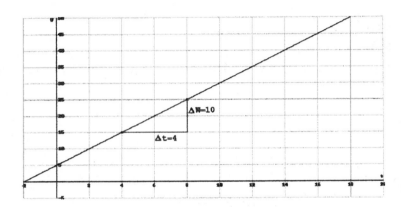

5. a.   The slope tells us the cost per pound of dry-roasted peanuts.

    b.   Use (40, 60) and (80, 120).

    $$m = \frac{\Delta C}{\Delta p} = \frac{120 - 60}{80 - 40} = \frac{60}{40} = 1.5 \text{ dollars per pound}$$

    c.

    $$m = \frac{\Delta C}{\Delta p}$$

    $$\Delta C = m \Delta p$$
    $$= 1.5(120 - 100)$$
    $$= 1.5(20)$$
    $$= 30$$

    The cost will increase $30.

7. a. $$m = \frac{\Delta y}{\Delta x} = \frac{4}{10} = \frac{2}{5}$$

    b.   It would be steeper since $\frac{3}{5} > \frac{2}{5}$.

9.   a.   Use (2, 650) and (6, 350).
$$m = \frac{\Delta d}{\Delta h} = \frac{350-650}{6-2} = \frac{300}{4} = -75 \text{ miles per hour}$$

b.   The slope tells us the speed of the train.

c.   The vertical intercept is (0, 800). This means that Roy started out 800 miles from home.

d.

$$m = \frac{\Delta d}{\Delta h} \quad \text{Use } (h, 0) \text{ and } (0, 800).$$

$$-75 = \frac{0-800}{h-0}$$

$$-75h = -800$$

$$h = \frac{800}{75} = 10\frac{2}{3}$$

The horizontal intercept is $\left(10\frac{2}{3}, 0\right)$. This means that Roy will be home in

$10\frac{2}{3}$ hours.

11.   a.   Use (2, 16) and (8, 4).
$$m = \frac{\Delta T}{\Delta m} = \frac{4-16}{8-2} = \frac{-12}{6} = -2 \text{ degrees per minute}$$

b.   The slope tells us the rate at which the temperature is lowered.

c.   The vertical intercept is (0, 20). This means that the compound was originally at 20°C.

d.   The horizontal intercept is (10, 0).   After 10 minutes, the compound will be at 0°C.

13.   a.   Use (−6, −5) and (2, −1).
$$m = \frac{\Delta y}{\Delta x} = \frac{-1-(-5)}{2-(-6)} = \frac{-1+5}{2+6} = \frac{4}{8} = \frac{1}{2}$$

b.

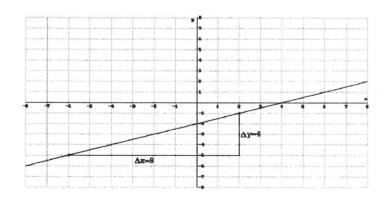

15.　a.　Use $(-3, -2)$ and $(6, -8)$.

$$m = \frac{\Delta y}{\Delta x} = \frac{-8-(-2)}{6-(-3)} = \frac{-8+2}{6+3} = \frac{-6}{9} = \frac{-2}{3}$$

　　b.

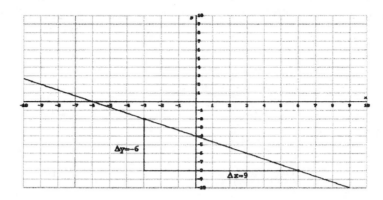

17.　a.

x	-2	0	2
y	-1	3	7

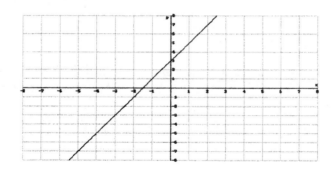

　　b.　Use $(0, 3)$ and $(2, 7)$.

$$m = \frac{\Delta y}{\Delta x} = \frac{7-3}{2-0} = \frac{4}{2} = 2$$

　　c.　$(0, 3)$

19.　a.

x	-4	0	4
y	-9	-6	-3

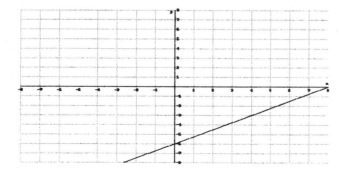

　　b.　Use $(0, -6)$ and $(4, -3)$.

$$m = \frac{\Delta y}{\Delta x} = \frac{-3-(-6)}{4-0} = \frac{3}{4}$$

　　c.　$(0, -6)$

Lesson 8.5

21. **a.**

x	-3	0	3
y	6	2	-2

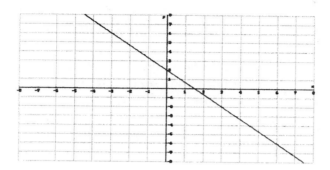

b. Use $(0, 2)$ and $(3, -2)$.

$$m = \frac{\Delta y}{\Delta x} = \frac{-2-2}{3-0} = \frac{-4}{3}$$

c. $(0, 2)$

23.

$$\frac{3}{2p^2q} + \frac{3}{8pq^2} = \frac{3 \cdot 4q}{2p^2q \cdot 4q} + \frac{3 \cdot p}{8pq^2 \cdot p}$$

$$= \frac{12q}{8p^2q^2} + \frac{3p}{8p^2q^2}$$

$$= \frac{12q + 3p}{8p^2q^2}$$

25.

$$\frac{g^2}{3h^2} - \frac{3}{g} = \frac{g^2 \cdot g}{3h^2 \cdot g} - \frac{3 \cdot 3h^2}{g \cdot 3h^2}$$

$$= \frac{g^3}{3h^2g} - \frac{9h^2}{3h^2g}$$

$$= \frac{g^3 - 9h^2}{3h^2g}$$

27.

$$b - \frac{2}{b} = \frac{b \cdot b}{1 \cdot b} - \frac{2}{b}$$

$$= \frac{b^2}{b} - \frac{2}{b}$$

$$= \frac{b^2 - 2}{b}$$

29.

$$1 - \frac{1}{x+1} = \frac{1 \cdot (x+1)}{1 \cdot (x+1)} - \frac{1}{x+1}$$

$$= \frac{x+1}{x+1} - \frac{1}{x+1}$$

$$= \frac{x+1-1}{x+1}$$

$$= \frac{x}{x+1}$$

31. 50% of $59.98 \approx \frac{1}{2}(60) = 30$

b

33. 78% of $793 \approx \frac{3}{4}(800) = 600$

c

35. 3.9% of $25{,}000 \approx \frac{1}{25}(25{,}000) = 1{,}000$

c

37. $\frac{17}{35} \times 100 \approx \frac{17}{34} \times 100 = 50\%$

c

39. $\frac{6.68}{99} \times 100 \approx \frac{7}{100} \times 100 = 7\%$

b

Chapter 8 Review

1.   100

3. P: part; r: percentage rate; W: whole

5.   increase; original amount

7.   If two variables are proportional, one will be a constant multiple of the other.

9.   Cross-multiplying is a technique used to solve proportions. You multiply the numerator of each ratio times the denominator of the other ratio, then set the products equal to each other. Cross-multiplying can only be used on proportions.

	Fraction	*Decimal*	*Percent*
11.	$\dfrac{1}{8}$	**0.125**	**12.5%**
13.	$\dfrac{8}{100} = \dfrac{2}{25}$	**0.08**	8%
15.	$\dfrac{35}{100} = \dfrac{7}{20}$	0.35	**35%**

17.
$$P = 0.15(13.00)$$
$$\text{discount} = \$1.95$$
$$\text{sales price} = 13.00 - 1.95$$
$$= \$11.05$$

19.
$$424.15 = 0.085W$$
$$\frac{424.15}{0.085} = \frac{0.085W}{0.085}$$
$$4990 = W$$
$$\text{price of computer} = \$4990$$
$$\text{total bill} = 4990 + 424.15$$
$$= \$5414.15$$

21.   a.   17.6%
      b.   Plastics
      c.   Paper
      d.

Material	*Percent of Waste*	*Millions of Tons of Waste*
Paper	40	$(0.40)(179.6) = 71.8$
Food Waste	7.3	$(0.073)(179.6) = 13.1$
Yard Waste	17.6	$(0.176)(179.6) = 31.6$
Metal	8.5	$(0.085)(179.6) = 15.3$
Glass	7	$(0.07)(179.6) = 12.6$
Plastic	8	$(0.08)(179.6) = 14.4$
Other	11.6	$(0.116)(179.6) = 20.8$

23.
$$\text{salary next year} = (0.90)(30,000) \qquad \text{salary following year} = (1.11)(27,000)$$
$$= \$27,000 \qquad\qquad\qquad = \$29,970$$

25. $\%\text{ increase for }25\text{-}34 = \dfrac{26{,}533 - 26{,}100}{26{,}100} \times 100 = \dfrac{433}{26{,}100} \times 100 \approx 1.66\%$

$\%\text{ increase for }65\text{ and older} = \dfrac{35{,}256 - 34{,}473}{34{,}473} \times 100 = \dfrac{783}{34{,}473} \times 100 \approx 2.27\%$

The 65 or older age group had the larger percentage increase.

27. a. $\begin{aligned}\text{starting price} &= 42.63 + 0.56 \\ &= \$43.19\end{aligned}$

   b. $\dfrac{-0.56}{43.19} \times 100 \approx -1.3\%$

29. $\begin{aligned}\text{ratio of fat to carbs} &= \dfrac{\text{grams of fat}}{\text{grams of carbs}} \\ &= \dfrac{65}{300} \\ &= \dfrac{13}{60} = 0.21\overline{6}\end{aligned}$

31. $\begin{aligned}\text{costumes per day} &= \dfrac{\text{number of costumes}}{\text{number of days}} \\ &= \dfrac{26}{4} = \dfrac{13}{2} \\ &= 6.5 \text{ costumes per day}\end{aligned}$

33. $\dfrac{w}{t} = \dfrac{3.6}{2} = 1.8; \quad \dfrac{w}{t} = \dfrac{9}{5} = 1.8$

$\dfrac{w}{t} = \dfrac{10.8}{6} = 1.8; \quad \dfrac{w}{t} = \dfrac{27}{15} = 1.8$

Yes; All the ratios are equal.

35. No; z is not a constant multiple of u.

37. No; The graph does not go through the origin.

39. a. $k = \dfrac{c}{p} = \dfrac{32}{4} = 8$

   $c = 8p$

   b.

p	c
4	32
5	40
10	80

   c. $c = 8p$

   $c = 8(10)$

   $c = 80$

   80 calories

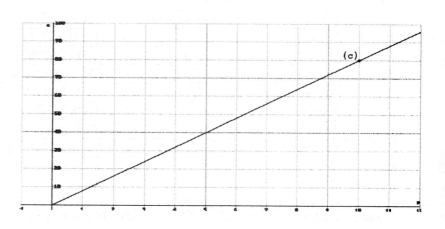

41.

$$\frac{18}{n} = \frac{15}{17.5}$$

$$18 \cdot 17.5 = n \cdot 15$$

$$315 = 15n$$

$$\frac{315}{15} = \frac{15n}{15}$$

$$21 = n$$

$$n = 21$$

43.

$$\frac{4}{x} = \frac{x}{9}$$

$$4 \cdot 9 = x \cdot x$$

$$36 = x^2$$

$$\pm\sqrt{36} = x$$

$$x = \pm 6$$

45. a. $\dfrac{3}{2}$

b.

$$\frac{3}{2} = \frac{18}{x}$$

$$2 \cdot 18 = 3x$$

$$36 = 3x$$

$$\frac{36}{3} = \frac{3x}{3}$$

$$12 = x$$

12 adults are needed.

c.

$$\frac{3}{2} = \frac{x}{16}$$

$$2 \cdot x = 3 \cdot 16$$

$$2x = 48$$

$$\frac{2x}{2} = \frac{48}{2}$$

$$x = 24$$

24 children can go in free.

47. a. $\dfrac{45}{300} = 0.15$

b.

$$\frac{45}{300} = \frac{105}{x}$$

$$45 \cdot x = 300 \cdot 105$$

$$45x = 31,500$$

$$\frac{45x}{45} = \frac{31,500}{45}$$

$$x = 700$$

She will have to walk 700 meters.

c. $\dfrac{1}{6} = 0.1\overline{6}$

The second hill is steeper.

49.

$$\frac{0.5}{6} = \frac{1.2}{x}$$

$$0.5 \cdot x = 6 \cdot 1.2$$

$$0.5x = 7.2$$

$$\frac{0.5x}{0.5} = \frac{7.2}{0.5}$$

$$x = 14.4$$

There are 14.4 grams of sugar in 1.2 cups.

51. I and V are similar.

a. $\dfrac{8}{4} = 2$

b. ratio of areas $= 2^2 = 4$

53. a.

$$\text{scale factor} = \frac{4}{1200} = \frac{1}{300}$$

$$900 \cdot \frac{1}{300} = 3$$

The width is 3 meters.

b. $30 \cdot \frac{1}{300} = 0.1$

The scale model should be 0.1 meters.

c.

$$\text{area} = \left(\frac{1}{300}\right)^2 60,000$$

$$= \frac{2}{3} \text{ square meters}$$

d.

$$\text{volume} = \left(\frac{1}{300}\right)^3 27,000$$

$$= 0.001 \text{ cubic meters}$$

55. a.

n	0	5	10
A	5	11	17

b. The vertical intercept tells us that Rani and Larry deposited $5,000 when Colby was born.

c. Use (0, 5) and (5, 11).

$$m = \frac{\Delta A}{\Delta n} = \frac{11-5}{5-0} = \frac{6}{5} = 1.2 \text{ thousand dollars per year}$$

d. The slope tells us that the college fund increases $1200 dollars per year.

57. a.

x	−4	0	4
y	−7	−2	3

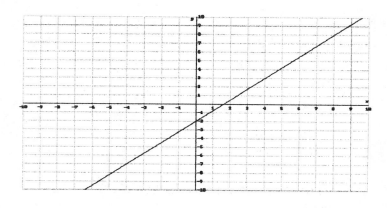

b. Use (0, −2) and (4, 3).

$$m = \frac{\Delta y}{\Delta x} = \frac{3-(-2)}{4-0} = \frac{5}{4}$$

59. Bearclaw Trail: $m = \dfrac{\Delta y}{\Delta x} = \dfrac{800}{3 \cdot 5280} = \dfrac{800}{15,840} \approx 0.0505$

Elkhorn Trail: $m = \dfrac{\Delta y}{\Delta x} = \dfrac{1200}{5 \cdot 5280} = \dfrac{1200}{26,400} \approx 0.0455$

Bearclaw Trail is steeper since it has a greater slope.

61. Total $\approx 2.20 + 6 + 3(1.50) = \$12.70$
No

63. $160.9 \div 100 = 1.609$

65. $250 \times 10 = 2500$ grams

67. $0.125(480) = \dfrac{1}{8}(480) = \dfrac{480}{8} = 60$

69. 15% of $800 = 80 + 40 = \$120$

71. 27% of $\$798 \approx \dfrac{1}{4}(800) = \$200$

73. 20% of $\$68.95 \approx \dfrac{1}{5}(70) = \$14$

b

Homework 9.1 A

1.

$$8x - 3x = 20$$
$$5x = 20$$
$$\frac{5x}{5} = \frac{20}{5}$$
$$x = 4$$

3.

$$-36 = 2z + 8 - 6z$$
$$-36 = -4z + 8$$
$$\underline{-8 \qquad -8}$$
$$-44 = -4z$$
$$\frac{-44}{-4} = \frac{-4z}{-4}$$
$$11 = z$$
$$z = 11$$

5.

$$-7a - 5a - 12 = 0$$
$$-12a - 12 = 0$$
$$\underline{+12 \; +12}$$
$$-12a = 12$$
$$\frac{-12a}{-12} = \frac{12}{-12}$$
$$a = -1$$

7.

$$16 - 4(9c) = 34$$
$$16 - 36c = 34$$
$$\underline{-16 \qquad -16}$$
$$-36c = 18$$
$$\frac{-36c}{-36} = \frac{18}{-36}$$
$$c = -0.5$$

9.

$$9 + 4f - 6f = 12$$
$$9 - 2f = 12$$
$$\underline{-9 \qquad -9}$$
$$-2f = 3$$
$$\frac{-2f}{-2} = \frac{3}{-2}$$
$$f = -1.5$$

11.

$$-7m + 17 + 3(2m) = 3$$
$$-7m + 17 + 6m = 3$$
$$-m + 17 = 3$$
$$\underline{-17 \; -17}$$
$$-m = -14$$
$$\frac{-m}{-1} = \frac{-14}{-1}$$
$$m = 14$$

13.

$$-24 = 26p + 2(15p) - 24$$
$$-24 = 26p + 30p - 24$$
$$-24 = 56p - 24$$
$$\underline{+24 \qquad \quad +24}$$
$$0 = 56p$$
$$\frac{0}{56} = \frac{56p}{56}$$
$$0 = p$$
$$p = 0$$

15.

$$-6s - s - 10 = 0$$
$$-7s - 10 = 0$$
$$\underline{+10 \; +10}$$
$$-7s = 10$$
$$\frac{-7s}{-7} = \frac{10}{-7}$$
$$s = -\frac{10}{7}$$

17.
$$-6s - s - 10s = 0$$
$$-17s = 0$$
$$\frac{-17s}{-17} = \frac{0}{-17}$$
$$s = 0$$

19.
$$-3(-2v) - 3(-2v) - 60 = 36$$
$$6v + 6v - 60 = 36$$
$$12v - 60 = 36$$
$$\underline{+60 \quad +60}$$
$$12v = 96$$
$$\frac{12v}{12} = \frac{96}{12}$$
$$v = 8$$

21. a. $3(6y) = 72$
     b.
$$3(6y) = 72$$
$$18y = 72$$
$$\frac{18y}{18} = \frac{72}{18}$$
$$y = 4$$
Each tape cost $4.

23. a. $3y + 6 = 72$
     b.
$$3y + 6 = 72$$
$$\underline{-6 \quad -6}$$
$$3y = 66$$
$$\frac{3y}{3} = \frac{66}{3}$$
$$y = 22$$
There are 22 cub scouts in each troop.

25. $42 \times 50 = \dfrac{42}{2} \times 100 = 21 \times 100 = 2100$

27. $28 \times 25 = \dfrac{28}{4} \times 100 = 7 \times 100 = 700$

29. $50 \times 72 = \dfrac{72}{2} \times 100 = 36 \times 100 = 3600$

31. $25 \times 48 = \dfrac{48}{4} \times 100 = 12 \times 100 = 1200$

33. $56 \times 25 = \dfrac{56}{4} \times 100 = 14 \times 100 = 1400$

Homework 9.1 B

1.
$$5z + 23 = 2z + 14$$
$$\underline{-2z \qquad -2z}$$
$$3z + 23 = 14$$
$$\underline{-23 \quad -23}$$
$$3z = -9$$
$$\frac{3z}{3} = \frac{-9}{3}$$
$$z = -3$$

3.
$$13 - 2b = 3b - 32$$
$$\underline{-3b \quad -3b}$$
$$-5b + 13 = -32$$
$$\underline{-13 \quad -13}$$
$$-5b = -45$$
$$\frac{-5b}{-5} = \frac{-45}{-5}$$
$$b = 9$$

5.

$$-2-3w+5w=-4w-44$$
$$-2+2w=-4w-44$$
$$\underline{+4w\quad+4w}$$
$$-2+6w=-44$$
$$\underline{+2\qquad\quad+2}$$
$$6w=-42$$
$$\frac{6w}{6}=\frac{-42}{6}$$
$$w=-7$$

7.

$$2x-7+3x=1+8x+4$$
$$5x-7=5+8x$$
$$\underline{-8x\qquad\quad-8x}$$
$$-3x-7=5$$
$$\underline{+7\ +7}$$
$$-3x=12$$
$$\frac{-3x}{-3}=\frac{12}{-3}$$
$$x=-4$$

9.

$$-12n-20=15n+4-3n$$
$$-12n-20=12n+4$$
$$\underline{+12n\qquad+12n}$$
$$-20=24n+4$$
$$\underline{-4\qquad\quad-4}$$
$$-24=24n$$
$$\frac{-24}{24}=\frac{24n}{24}$$
$$-1=n$$
$$n=-1$$

11. a. Ounces of catsup per batch: c
     b. Catsup used: $8c$
     $$5c+12$$
     $$8c=5c+12$$
     c.
     $$8c=5c+12$$
     $$\underline{-5c\ -5c}$$
     $$3c=12$$
     $$\frac{3c}{3}=\frac{12}{3}$$
     $$c=4$$

He uses 4 ounces of catsup in each batch of barbecue sauce.

13. a. Length of mailbox post: m
     b. Length of each wooden beam:
     $$3m+33$$
     $$m+131$$
     $$3m+33=m+131$$
     c.
     $$3m+33=m+131$$
     $$\underline{-m\qquad-m}$$
     $$2m+33=131$$
     $$\underline{-33\ -33}$$
     $$2m=98$$
     $$\frac{2m}{2}=\frac{98}{2}$$
     $$m=49$$

Each post is 49 inches long.

15. a. Cost of one can of cat food: c
     b. Checkout totals: $12c+1.80$
     $$8c+5$$
     $$12c+1.80=8c+5$$
     c.
     $$12c+1.80=8c+5$$
     $$\underline{-8c\qquad\quad-8c}$$
     $$4c+1.80=5$$
     $$\underline{-1.80\ -1.80}$$
     $$4c=3.20$$
     $$\frac{4c}{4}=\frac{3.20}{4}$$
     $$c=0.80$$

One can of cat food costs $0.80.

17. a. Milligrams of calcium in glass of milk: m

b. Daily requirement: $3m - 70$
$$2m + 220$$
$$3m - 70 = 2m + 220$$

c.
$$3m - 70 = 2m + 220$$
$$\underline{-2m \qquad -2m}$$
$$m - 70 = 220$$
$$\underline{+70 \; +70}$$
$$m = 290$$

A glass of milk contains 290 mg of calcium.

19. $15 + 39 = (15 + 40) - 1 = (55) - 1 = 54$

21. $45 + 28 = (45 + 30) - 2 = (75) - 2 = 73$

23. $59 + 34 = (34 + 60) - 1 = (94) - 1 = 93$

25. $48 + 27 = (27 + 50) - 2 = (77) - 2 = 75$

27. $37 + 26 = (26 + 40) - 3 = (66) - 3 = 63$

Homework 9.2 A

1. $3(x-4) = 3(x) - 3(4) = 3x - 12$

3. $5(2y-3) = 5(2y) - 5(3) = 10y - 15$

5.
$$-2(4x+8) = -2(4x) - 2(8)$$
$$= -8x - 16$$

7.
$$-5(4-5a) = -5(4) - 5(-5a)$$
$$= -20 + 25a$$

9.
$$-(5b-3) = -1(5b-3)$$
$$= -1(5b) - 1(-3)$$
$$= -5b + 3$$

11.
$$(-6+2t)(-6) = -6(-6) + 2t(-6)$$
$$= 36 - 12t$$

13. Let $b = 4$.
$$2(b+3) = 2(4+3) = 2(7) = 14$$
$$2b + 6 = 2(4) + 6 = 8 + 6 = 14$$
Yes

15. Let $z = 3$.
$$-5(2z-1) = -5[2(3)-1]$$
$$= -5[6-1]$$
$$= -5[5]$$
$$= -25$$
$$-10z + 5 = -10(3) + 5 = -30 + 5 = -25$$
Yes

17. Let $k = 1$.
$$(6-4k)(3) = [6-4(1)](3)$$
$$= [6-4](3)$$
$$= [2](3)$$
$$= 6$$
$$18 - 12k = 18 - 12(1) = 18 - 12 = 6$$
Yes

19. a.   Jim made $0.55(4) = \$2.20$.
Nora made $0.55(6) = \$3.30$.
Total: $\$2.20 + \$3.30 = \$5.50$
  b.   They collected $4 + 6 = 10$ lbs.
Total: $0.55(10) = \$5.50$

21. a.
$$3(2x+5x) = 3(2x) + 3(5x)$$
$$= 6x + 15x$$
$$= 21x$$
  b. $3(2x+5x) = 3(7x) = 21x$

23.
$$-6(x+1) + 2x = -6(x) - 6(1) + 2x$$
$$= -6x - 6 + 2x$$
$$= -4x - 6$$

25.
$$4x - 9(2-3x) = 4x - 9(2) - 9(-3x)$$
$$= 4x - 18 + 27x$$
$$= 31x - 18$$

27.
$$5 - 2(4x-9) + 9x$$
$$= 5 - 2(4x) - 2(-9) + 9x$$
$$= 5 - 8x + 18 + 9x$$
$$= 23 + x$$

29.

$5(y+3) - 2(y-3)$

$= 5(y) + 5(3) - 2(y) - 2(-3)$

$= 5y + 15 - 2y + 6$

$= 3y + 21$

31.

$-3(2t - 5) + 2t - 5(t+2)$

$= -3(2t) - 3(-5) + 2t - 5(t) - 5(2)$

$= -6t + 15 + 2t - 5t - 10$

$= -9t + 5$

33. a. $5b$ means 5 multiplied times b. $5 + b$ means 5 added to b. The first expression is a product, and the second expression is a sum.

b. $2(5b)$ means 2 times $5b$. $2(5 + b)$ means 2 times the quantity $5 + b$.
In the second expression, 2 is multiplied times 5 and times b. The second expression requires the distributive law, while the first expression does not.

c. $2(5b) = 10b$; $2(5 + b) = 2(5) + 2(b) = 10 + 2b$

35. a. True
b. True
c. True
d. False

37. Since 548 is even, 548 is divisible by 2. Since 48 is divisible by 4, 548 is divisible by 4.

39. Since 2357 is odd, 2357 is not divisible by 2 nor by 4.

41. Since 4528 is even, 4528 is divisible by 2. Since 28 is divisible by 4, 4528 is divisible by 4.

43. Since 8762 is even, 8762 is divisible by 2. Since 62 is not divisible by 4, 8762 is not divisible by 4.

45. Since 1004 is even, 1004 is divisible by 2. Since 4 is divisible by 4, 1004 is divisible by 4.

Homework 9.2 B

1.   a.

$$2(x - 5) = 6$$
$$\frac{2(x - 5)}{2} = \frac{6}{2}$$
$$x - 5 = 3$$
$$\underline{+5 \quad +5}$$
$$x = 8$$

   b.

$$2(x - 5) = 6$$
$$2x - 10 = 6$$
$$\underline{+10 \quad +10}$$
$$2x = 16$$
$$\frac{2x}{2} = \frac{16}{2}$$
$$x = 8$$

3.

$$-4(x - 6) = 8$$
$$-4x + 24 = 8$$
$$\underline{-24 \quad -24}$$
$$-4x = -16$$
$$\frac{-4x}{-4} = \frac{-16}{-4}$$
$$x = 4$$

5.

$$5 - (2b + 9) = 16$$
$$5 - 2b - 9 = 16$$
$$-2b - 4 = 16$$
$$\underline{+4 \quad +4}$$
$$-2b = 20$$
$$\frac{-2b}{-2} = \frac{20}{-2}$$
$$b = -10$$

7.

$$-15 = 3(2 + 7x)$$
$$-15 = 6 + 21x$$
$$\underline{-6 \quad -6}$$
$$-21 = 21x$$
$$\frac{-21}{21} = \frac{21x}{21}$$
$$-1 = x$$
$$x = -1$$

9.

$$14 + 8(3z - 5) = 34$$
$$14 + 24z - 40 = 34$$
$$24z - 26 = 34$$
$$\underline{+26 \quad +26}$$
$$24z = 60$$
$$\frac{24z}{24} = \frac{60}{24}$$
$$z = \frac{5}{2}$$

11.

$$-64 = 2(1 - 3w) - 5w$$
$$-64 = 2 - 6w - 5w$$
$$-64 = 2 - 11w$$
$$\underline{-2 \quad -2}$$
$$-66 = -11w$$
$$\frac{-66}{-11} = \frac{-11w}{-11}$$
$$6 = w$$
$$w = 6$$

13.

$$9 = 6p - (p - 4) - 20$$
$$9 = 6p - p + 4 - 20$$
$$9 = 5p - 16$$
$$\underline{+16 \qquad +16}$$
$$25 = 5p$$
$$\frac{25}{5} = \frac{5p}{5}$$
$$5 = p$$
$$p = 5$$

15.

$$-30c - 30 + 5(3c + 4) = -10$$
$$-30c - 30 + 15c + 20 = -10$$
$$-15c - 10 = -10$$
$$\underline{+10 \quad +10}$$
$$-15c = 0$$
$$\frac{-15c}{-15} = \frac{0}{-15}$$
$$c = 0$$

17.

$$0.6b + 0.3(b - 6) = 9$$
$$0.6b + 0.3b - 1.8 = 9$$
$$0.9b - 1.8 = 9$$
$$\underline{+1.8 \quad +1.8}$$
$$0.9b = 10.8$$
$$\frac{0.9b}{0.9} = \frac{10.8}{0.9}$$
$$b = 12$$

19.

$$0.7(d + 20) - 0.5(d + 30) = 13$$
$$0.7d + 14 - 0.5d - 15 = 13$$
$$0.2d - 1 = 13$$
$$\underline{+1 \quad +1}$$
$$0.2d = 14$$
$$\frac{0.2d}{0.2} = \frac{14}{0.2}$$
$$d = 70$$

21. $72 + 6h$

23. $300 - 6d$

25. $50 + \dfrac{P}{12}$

27. $\dfrac{S - 60}{15}$

29. $20 + 0.02B$

31. $4 + 5 + 8 = 17$
Since 17 is not divisible by 3, 458 is not divisible by 3.
Since 17 is not divisible by 9, 458 is not divisible by 9.

33. $7 + 1 + 4 = 12$
Since 12 is divisible by 3, 714 is divisible by 3.
Since 12 is not divisible by 9, 714 is not divisible by 9.

35. $1 + 1 + 1 + 6 = 9$
Since 9 is divisible by 3, 1116 is divisible by 3.
Since 9 is divisible by 9, 1116 is divisible by 9.

37. $8 + 1 + 9 + 7 = 25$
Since 25 is not divisible by 3, 8197 is not divisible by 3.
Since 25 is not divisible by 9, 8197 is not divisible by 9.

39. $9 + 0 + 1 + 8 = 18$
Since 18 is divisible by 3, 9018 is divisible by 3.
Since 18 is divisible by 9, 9018 is divisible by 9.

Homework 9.3 A

1. a. Number of pages in English assignment: x
   b. $x + 12$
   c. $x + (x + 12) = 2x + 12$

3. a. Price of jogging shoes: p
   b. $p - 20$
   c. $p + (p - 20) = 2p - 20$

5. a. Number of dogs treated: d
   b. $2d$
   c. $(d + 2d) - 5 = 3d - 5$

7. a. Number of children: c
   b. $24 - c$
   c. $24 - 2c$

9. a. Ounces of ice cream: c
   b. $10 - c$
   c. $0.12\,c$
   d. $0.55(10 - c)$
   e.
   $$0.12c + 0.55(10 - c)$$
   $$= 0.12c + 5.5 - 0.55c$$
   $$= 5.5 - 0.43c$$

11.
$$4u - 2 - 7u = 10$$
$$-3u - 2 = 10$$
$$\underline{+2 \quad +2}$$
$$-3u = 12$$
$$\frac{-3u}{-3} = \frac{12}{-3}$$
$$u = -4$$

13.
$$-2(-5h) - 8 = -20$$
$$10h - 8 = -20$$
$$\underline{+8 \quad +8}$$
$$10h = -12$$
$$\frac{10h}{10} = \frac{-12}{10}$$
$$h = -\frac{6}{5}$$

15.
$$-15 = 5(3x - 4)$$
$$-15 = 15x - 20$$
$$\underline{+20 \qquad +20}$$
$$5 = 15x$$
$$\frac{5}{15} = \frac{15x}{15}$$
$$\frac{1}{3} = x$$
$$x = \frac{1}{3}$$

17.
$$5 + 3(-2b + 4) = 17$$
$$5 - 6b + 12 = 17$$
$$-6b + 17 = 17$$
$$\underline{-17 \quad -17}$$
$$-6b = 0$$
$$\frac{-6b}{-6} = \frac{0}{-6}$$
$$b = 0$$

19.
$$-5 = 12 - (3M - 1)$$
$$-5 = 12 - 3M + 1$$
$$-5 = 13 - 3M$$
$$\underline{-13 \quad -13}$$
$$-18 = -3M$$
$$\frac{-18}{-3} = \frac{-3M}{-3}$$
$$6 = M$$
$$M = 6$$

21. Since the last digit is 0, 170 is divisible by 5 and 10.

23. Since the last digit is 5, 4355 is divisible by 5 but not by 10.

25. Since the last digit is 0, 5050 is divisible by 5 and 10.

27. Since the last digit is 1, 1001 is not divisible by 5 nor by 10.

29. Since the last digit is 5, 9015 is divisible by 5 but not by 10.

Homework 9.3 B

1. Step 1: Number of envelopes Ralph stuffed: x
   Number of envelopes Wanda stuffed: $x + 57$
   Step 2: Number of envelopes stuffed together: $x + (x + 57) = 349$
   Step 3:
$$x + (x + 57) = 349$$
$$2x + 57 = 349$$
$$\underline{-57 \quad -57}$$
$$2x = 292$$
$$\frac{2x}{2} = \frac{292}{2}$$
$$x = 146$$
   Wanda stuffed 146 envelopes. Ralph stuffed $146 + 57 = 203$ envelopes.

3. Step 1: Number of tables in smoking: x
   Number of tables in non-smoking: $2x$
   Step 2: Total number of tables: $x + 2x = 54$
   Step 3:
$$x + 2x = 54$$
$$3x = 54$$
$$\frac{3x}{3} = \frac{54}{3}$$
$$x = 18$$
   There are 18 tables in the smoking section.

5.   Step 1: Number of boys: x
                 Number of girls: $x - 6$
        Step 2: Number of children on the bus: $x + (x - 6) = 64$
        Step 3:

$$x + (x - 6) = 64$$
$$2x - 6 = 64$$
$$\underline{+6 \quad +6}$$
$$2x = 70$$
$$\frac{2x}{2} = \frac{70}{2}$$
$$x = 35$$

        There are 35 boys on the bus.

7.   Step 1: Cost of a child's ticket: x
                 Cost of an adult's ticket: $32 - x$
        Step 2: Cost of 4 child and 2 adult tickets: $4x + 2(32 - x) = 82$
        Step 3:

$$4x + 2(32 - x) = 82$$
$$4x + 64 - 2x = 82$$
$$2x + 64 = 82$$
$$\underline{-64 \quad -64}$$
$$2x = 18$$
$$\frac{2x}{2} = \frac{18}{2}$$
$$x = 9$$

        A child's ticket is \$9.

9.   Step 1: Length of shorter base: x
                 Length of longer base: $x + 6$
        Step 2: Area of table: $\frac{1}{2}(4)(x + x + 6) = 28$
        Step 3:

$$\frac{1}{2}(4)(x + x + 6) = 28$$
$$2(2x + 6) = 28$$
$$4x + 12 = 28$$
$$\underline{-12 \quad -12}$$
$$4x = 16$$
$$\frac{4x}{4} = \frac{16}{4}$$
$$x = 4$$

        The shorter base is 4 feet long. The longer base is $4 + 6 = 10$ feet long.

11. a.   Number of coach tickets: x
       Number of first-class tickets: $140 - x$

   b.   Receipts from coach tickets: $150x$
       Receipts from first-class tickets: $230(140 - x)$

   c. $150x + 230(140 - x) = 22,600$

   d.

$$150x + 230(140 - x) = 22,600$$
$$150x + 32,200 - 230x = 22,600$$
$$-80x + 32,200 = 22,600$$
$$\underline{-32,200 \quad -32,200}$$
$$-80x = -9,600$$
$$\frac{-80x}{-80} = \frac{-9,600}{-80}$$
$$x = 120$$

120 coach tickets were sold. $140 - 120 = 20$ first-class tickets were sold.

13.
$$7(y - 5) + 3(5 - y) = 7y - 35 + 15 - 3y$$
$$= 4y - 20$$

15.
$$6 - 2(a - 3) - 4(1 + 3a) + 12 - 2 = 6 - 2a + 6 - 4 - 12a + 12 - 2$$
$$= -14a + 18$$

17.
$$24 - 8n + 4 = -6 - 10n$$
$$28 - 8n = -6 - 10n$$
$$\underline{+10n \qquad +10n}$$
$$28 + 2n = -6$$
$$\underline{-28 \qquad -28}$$
$$2n = -34$$
$$\frac{2n}{2} = \frac{-34}{2}$$
$$n = -17$$

19.
$$0.30(h + 200) + 0.50h = 0.60h + 220$$
$$0.30h + 60 + 0.50h = 0.60h + 220$$
$$0.80h + 60 = 0.60h + 220$$
$$\underline{-0.60h \qquad -0.60h}$$
$$0.20h + 60 = 220$$
$$\underline{-60 \quad -60}$$
$$0.20h = 160$$
$$\frac{0.20h}{0.20} = \frac{160}{0.20}$$
$$h = 800$$

21.
$$44 - 8 = (30 + 14) - 8$$
$$= 30 + (14 - 8)$$
$$= 30 + 6$$
$$= 36$$

23.
$$73 - 6 = (60 + 13) - 6$$
$$= 60 + (13 - 6)$$
$$= 60 + 7$$
$$= 67$$

25.

$$96 - 7 = (80 + 16) - 7$$
$$= 80 + (16 - 7)$$
$$= 80 + 9$$
$$= 89$$

27.

$$54 - 5 = (40 + 14) - 5$$
$$= 40 + (14 - 5)$$
$$= 40 + 9$$
$$= 49$$

29.

$$68 - 9 = (50 + 18) - 9$$
$$= 50 + (18 - 9)$$
$$= 50 + 9$$
$$= 59$$

Homework 9.4

1. the LCD

3.

$$\frac{1}{6}m + 3 = \frac{-5}{6}$$

$$6\left(\frac{1}{6}m + 3\right) = \left(\frac{-5}{6}\right)6$$

$$6\left(\frac{1}{6}m\right) + 6(3) = \left(\frac{-5}{6}\right)6$$

$$m + 18 = -5$$

$$m = -23$$

5.

$$\frac{2}{3}c + \frac{4}{3} = 2c + 4$$

$$3\left(\frac{2}{3}c + \frac{4}{3}\right) = (2c + 4)3$$

$$3\left(\frac{2}{3}c\right) + 3\left(\frac{4}{3}\right) = (2c)3 + 4(3)$$

$$2c + 4 = 6c + 12$$

$$-4c + 4 = 12$$

$$-4c = 8$$

$$c = -2$$

7.

$$\frac{2}{3}s - 6 = \frac{8}{3}s - 3$$

$$3\left(\frac{2}{3}s - 6\right) = \left(\frac{8}{3}s - 3\right)3$$

$$3\left(\frac{2}{3}s\right) - 3(6) = \left(\frac{8}{3}s\right)3 - (3)3$$

$$2s - 18 = 8s - 9$$

$$-6s = 9$$

$$s = \frac{9}{-6} = -\frac{3}{2}$$

9.

$$z + \frac{1}{7} = \frac{2}{3}z$$

$$21\left(z + \frac{1}{7}\right) = \left(\frac{2}{3}z\right)21$$

$$21(z) + 21\left(\frac{1}{7}\right) = \left(\frac{2}{3}z\right)21$$

$$21z + 3 = 14z$$

$$3 = -7z$$

$$-\frac{3}{7} = z$$

$$z = -\frac{3}{7}$$

11.

$$\frac{1}{6}p - \frac{7}{3} = \frac{2}{9}p - \frac{1}{4}p$$

$$36\left(\frac{1}{6}p - \frac{7}{3}\right) = \left(\frac{2}{9}p - \frac{1}{4}p\right)36$$

$$36\left(\frac{1}{6}p\right) - 36\left(\frac{7}{3}\right) = \left(\frac{2}{9}p\right)36 - \left(\frac{1}{4}p\right)36$$

$$6p - 84 = 8p - 9p$$

$$6p - 84 = -p$$

$$-84 = -7p$$

$$12 = p$$

$$p = 12$$

13. Length of the garden: x

$$1\frac{3}{8}x = 7\frac{1}{3}$$

$$\frac{11}{8}x = \frac{22}{3}$$

$$24\left(\frac{11}{8}x\right) = \left(\frac{22}{3}\right)24$$

$$33x = 176$$

$$x = \frac{176}{33} = \frac{16}{3} = 5\frac{1}{3}$$

The garden is $5\frac{1}{3}$ yards long.

15. Number of pages: x

$$\frac{2}{5}x + \frac{1}{4}x + 308 = x$$

$$20\left(\frac{2}{5}x + \frac{1}{4}x + 308\right) = (x)20$$

$$20\left(\frac{2}{5}x\right) + 20\left(\frac{1}{4}x\right) + 20(308) = (x)20$$

$$8x + 5x + 6160 = 20x$$

$$13x + 6160 = 20x$$

$$6160 = 7x$$

$$x = 880$$

There are 880 pages in *Don Quixote*.

17. a. Number of bakers: b
    b. $14 - b$

19. a. $150b$
    b. $100(14 - b)$
    c.

$$150b + 100(14 - b)$$
$$= 150b + 1400 - 100b$$
$$= 50b + 1400$$

21. a. Number of oak trees: k
    b. $3k$

23. a. $20k$
    b. $9(3k)$
    c.

$$20k + 9(3k) = 20k + 27k$$
$$= 47k$$

25. a. Number of students enrolled in algebra course: c
    b. $840 - c$
    c. $0.40c$
    d. $0.10(840 - c)$
    e.

$$0.40c + 0.10(840 - c)$$
$$= 0.40c + 84 - 0.10c$$
$$= 0.30c + 84$$

27. Number of apples in basket: x

$$\frac{1}{3}x + \frac{1}{8}x + \frac{1}{4}x + \frac{1}{5}x + 10 + 1 = x$$

$$120\left(\frac{1}{3}x + \frac{1}{8}x + \frac{1}{4}x + \frac{1}{5}x + 10 + 1\right) = (x)120$$

$$120\left(\frac{1}{3}x\right) + 120\left(\frac{1}{8}x\right) + 120\left(\frac{1}{4}x\right) + 120\left(\frac{1}{5}x\right) + 120(11) = (x)120$$

$$40x + 15x + 30x + 24x + 1320 = 120x$$

$$109x + 1320 = 120x$$

$$1320 = 11x$$

$$x = 120$$

There were 120 apples in the basket originally.

29.
$$81 - 5 = (80 + 1) - 5$$
$$= (80 - 5) + 1$$
$$= 75 + 1$$
$$= 76$$

31.
$$62 - 9 = (60 + 2) - 9$$
$$= (60 - 9) + 2$$
$$= 51 + 2$$
$$= 53$$

33.
$$36 - 9 = (30 + 6) - 9$$
$$= (30 - 9) + 6$$
$$= 21 + 6$$
$$= 27$$

35.
$$93 - 4 = (90 + 3) - 4$$
$$= (90 - 4) + 3$$
$$= 86 + 3$$
$$= 89$$

37.
$$44 - 6 = (40 + 4) - 6$$
$$= (40 - 6) + 4$$
$$= 34 + 4$$
$$= 38$$

Homework 9.5 A

1. a. $I = Prt$
   $$= x(0.05)(1)$$
   $$= 0.05x$$

   b.
   $$I = Prt$$
   $$140 = x(0.05)(1)$$
   $$140 = 0.05x$$
   $$\frac{140}{0.05} = \frac{0.05x}{0.05}$$
   $$2800 = x$$
   You invested $2800.

3. $I = Prt$
   $$25{,}000 = (50{,}000)r(4)$$
   $$25{,}000 = 200{,}000r$$
   $$\frac{25{,}000}{200{,}000} = \frac{200{,}000r}{200{,}000}$$
   $$0.125 = r$$
   He should charge 12.5% interest.

5. Interest from credit union:
   $$I = Prt$$
   $$= (1200)(0.065)(5)$$
   $$= \$390$$
   Interest from city bond:
   $$I = Prt$$
   $$= (600)(0.08)(5)$$
   $$= \$240$$
   Total interest: $390 + 240 = \$630$

7. a. Amount borrowed from city fund: x
   Amount borrowed from savings and loan: $25{,}000 - x$

   b. Interest owed on loan from city fund:
   $$I = Prt$$
   $$= x(0.05)(1)$$
   $$= 0.05x$$
   Interest owed on loan from savings and loan:
   $$I = Prt$$
   $$= (25{,}000 - x)(0.084)(1)$$
   $$= 0.084(25{,}000 - x)$$

   c. Total interest owed after one year:
   $$0.05x + 0.084(25{,}000 - x) = 0.05x + 2100 - 0.084x$$
   $$= 2100 - 0.034x$$

9. a.

Amount invested in savings	Amount invested in T-bill	Interest from savings	Interest from T-bill	Total interest
$100	$1200-$100 =$1100	$100(0.05)= $5	$1100(0.075)= $82.50	$5+$82.50= $87.50
$200	$1000	$10	$75	$85
$300	$900	$15	$67.50	$82.50
$400	$800	$20	$60	$80
$500	$700	$25	$52.50	$77.50
$600	$600	$30	$45	$75
$700	$500	$35	$37.50	$72.50
$800	$400	$40	$30	$70
$900	$300	$45	$22.50	$67.50
$1000	$200	$50	$15	$65
$1100	$100	$55	$7.50	$62.50

b. Amount invested in savings account: x
Amount invested in T-bill: $1200 - x$
Interest from savings account: $0.05x$
Interest from T-bill: $0.075(1200 - x)$
Total interest:

$$0.05x + 0.075(1200 - x) = 0.05x + 90 - 0.075x$$
$$= 90 - 0.025x$$

11. a. $5000 - x$
b. Interest earned on stocks: Interest earned on bonds:
$$I = Prt$$ $$I = Prt$$
$$= x(0.12)(1)$$ $$= (5000 - x)(0.07)(1)$$
$$= 0.12x$$ $$= 0.07(5000 - x)$$

Total interest earned:
$$I = 0.12x + 0.07(5000 - x)$$
$$= 0.12x + 350 - 0.07x$$
$$= 0.05x + 350$$

c.

$$440 = 0.05x + 350$$
$$\underline{-350 \qquad\qquad -350}$$
$$90 = 0.05x$$
$$\frac{90}{0.05} = \frac{0.05x}{0.05}$$
$$1800 = x$$

You invested $1800 in stocks and $5000 - 1800 = 3200 in bonds.

13.   Amount invested in video-pizza: x
       Amount invested in T-bill: $x + 6000$

$$334.50 = -0.096x + 0.073(x + 6000)$$
$$334.50 = -0.096x + 0.073x + 438$$
$$334.50 = -0.023x + 438$$
$$\underline{-438 \qquad\qquad -438}$$
$$-103.5 = -0.023x$$
$$4500 = x$$

She invested $4500 in the video-pizza venture.

15.

$$44 - 8 = 44 - (4 + 4)$$
$$= (44 - 4) - 4$$
$$= 40 - 4$$
$$= 36$$

17.

$$73 - 26 = 73 - (23 + 3)$$
$$= (73 - 23) - 3$$
$$= 50 - 3$$
$$= 47$$

19.

$$96 - 57 = 96 - (56 + 1)$$
$$= (96 - 56) - 1$$
$$= 40 - 1$$
$$= 39$$

21.

$$54 - 35 = 54 - (34 + 1)$$
$$= (54 - 34) - 1$$
$$= 20 - 1$$
$$= 19$$

23.

$$68 - 19 = 68 - (18 + 1)$$
$$= (68 - 18) - 1$$
$$= 50 - 1$$
$$= 49$$

Homework 9.5 B

1.   a.

$$P = rW$$
$$= 0.30(8)$$
$$= 2.4 \text{ gallons}$$

      b.

$$P = rW$$
$$= 0.30x \text{ gallons}$$

3.

	Number of m&m's	% Red	Number of Red m&m's
First bag	50	0.30	15
Second bag	30	0.70	21
Mixture	80	0.45	36

   a.   First bag: Number red $= rW = 0.30(50) = 15$

             Second bag: Number red $= rW = 0.70(30) = 21$

   b.   Total number of m&m's: $50 + 30 = 80$

             Number of red: $15 + 21 = 36$

   c.   % red in mixture $= \dfrac{P}{W} = \dfrac{36}{80} = 0.45$

             45% of the mixture is red.

   d.   Number red $= rW = 0.45(80) = 36$

5.   a.   20% solution: acid $= rW = 0.20(5) = 1$ quart

             30% solution: acid $= rW = 0.30(10) = 3$ quarts

      b.   Amount of acid: $1 + 3 = 4$ quarts

             Amount of mixture: $5 + 10 = 15$ quarts

      c.   % acid $= \dfrac{P}{W} = \dfrac{4}{15} = 0.2\overline{6}$

             The mixture will be $26.\overline{6}\%$ acid.

      d.

	Quarts of solution	Strength (% acid)	Quarts of acid
20% solution	5	0.20	1
30% solution	10	0.30	3
Mixture	15	$0.2\overline{6}$	4

7.   a.   Quarts of 30% solution: $15 - x$

      b.

	Quarts of solution	Strength (% acid)	Quarts of acid
20% solution	x	0.20	$0.20x$
30% solution	$15 - x$	0.30	$0.30(15 - x)$
Mixture	15	0.28	4.2

9.

	Number produced	% Meeting standards	Number meeting standards
Cars	50,000	0.30	15,000
Trucks	x	0.80	$0.80x$
All vehicles	$50,000 + x$	0.60	$0.60(50,000 + x)$

a.   Number of trucks: x

b.   See table.

c.   See table.

d.   Number of cars meeting standards: $P = rW = 0.30(50,000) = 15,000$

   Number of trucks meeting standards: $P = rW = 0.80(x) = 0.80x$

e.   Total number of vehicles meeting standards:

$$\left(\begin{array}{c}\text{Number of cars}\\\text{meeting standards}\end{array}\right) + \left(\begin{array}{c}\text{Number of trucks}\\\text{meeting standards}\end{array}\right) = 15,000 + 0.80x$$

$$\left(\begin{array}{c}\text{Total number}\\\text{produced}\end{array}\right)\left(\begin{array}{c}\text{\% meeting}\\\text{standards}\end{array}\right) = 0.60(50,000 + x)$$

$$15,000 + 0.80x = 0.60(50,000 + x)$$

f.

$$15,000 + 0.80x = 0.60(50,000 + x)$$
$$15,000 + 0.80x = 30,000 + 0.60x$$
$$\underline{\quad -0.60x \qquad\qquad -0.60x\quad}$$
$$15,000 + 0.20x = 30,000$$
$$\underline{-15,000 \qquad\quad -15,000\quad}$$
$$0.20x = 15,000$$
$$\frac{0.20x}{0.20} = \frac{15,000}{0.20}$$
$$x = 75,000$$

They should produce 75,000 trucks.

11.
$$8 + 29 + 32 = (8 + 32) + 29$$
$$= 40 + 29$$
$$= 69$$

13.
$$14 + 3 + 27 + 6 = (14 + 6) + (3 + 27)$$
$$= 20 + 30$$
$$= 50$$

15.
$$26 + 32 + 4 + 8 = (26 + 4) + (32 + 8)$$
$$= 30 + 40$$
$$= 70$$

17.
$$25 + 6 + 25 + 34 = (25 + 25) + (6 + 34)$$
$$= 50 + 40$$
$$= 90$$

19.
$$17+36+24+33 = (17+33)+(36+24)$$
$$= 50+60$$
$$=110$$

Homework 9.5 C

1. a.   Wendy's distance: Carl's distance:
       $d = rt$ $d = rt$
       $= (60)(2)$ $= (45)(2)$
       $= 120 \text{ miles}$ $= 90 \text{ miles}$
   Distance between Wendy and Carl: $120 + 90 = 210 \text{ miles}$

   b.

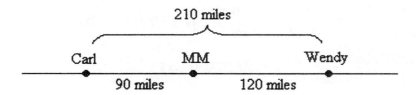

3. a.   Rate upstream = cruising speed – current = $21 - 3 = 18$ miles per hour
       $d = rt$
       $12 = 18t$
       $\dfrac{12}{18} = \dfrac{18t}{18}$
       $\dfrac{2}{3} = t$

   The trip upstream is $\dfrac{2}{3}$ hour, or 40 minutes.

   b.   Rate downstream = cruising speed + current = $21 + 3 = 24$ miles per hour
       $d = rt$
       $12 = 24t$
       $\dfrac{12}{24} = \dfrac{24t}{24}$
       $\dfrac{1}{2} = t$

   The trip upstream is $\dfrac{1}{2}$ hour, or 30 minutes.

5. First part of trip:
    $$d = rt$$
    $$60 = 30t$$
    $$\frac{60}{30} = \frac{30t}{30}$$
    $$2 = t$$

    Second part of trip:
    $$d = rt$$
    $$60 = 50t$$
    $$\frac{60}{50} = \frac{50t}{50}$$
    $$1.2 = t$$

    Whole trip: $t = 2 + 1.2 = 3.2$
    $$d = rt$$
    $$120 = r(3.2)$$
    $$\frac{120}{3.2} = \frac{r(3.2)}{3.2}$$
    $$37.5 = r$$

    Her average speed for the whole trip was 37.5 miles per hour.

7. a. $w + 2$
    b.   Claire: Brooke:
          $$d = rt$$ $$d = rt$$
          $$= w(4)$$ $$= (w + 2)(4)$$
          $$= 4w$$ $$= 4(w + 2)$$
    c.

d = 4(w+2)   Brooke   d = 4w   Claire

9. a. $h - 3$
    b.   Oscar: Felix:
          $$d = rt$$ $$d = rt$$
          $$= 45h$$ $$= 70(h - 3)$$
    c.

    Oscar: $d = 45h$

    Westbend   Eastbank

    Felix: $d = 70(h-3)$

11. **a.**

	Rate	Time	Distance
Delbert	*s*	6	6*s*
Francine	*s* − 5	6	6(*s* − 5)

b. See table.
c. See table.
d.

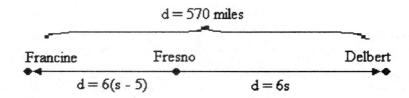

e. Distance apart after 6 hours:
$$6s + 6(s - 5) = 570$$

f.

$$6s + 6(s - 5) = 570$$
$$6s + 6s - 30 = 570$$
$$12s - 30 = 570$$
$$\underline{+30 \quad +30}$$
$$12s = 600$$
$$\frac{12s}{12} = \frac{600}{12}$$
$$s = 50$$

Delbert's speed was 50 mph, and Francine's speed was 50 − 5 = 45 mph.

13. a. Speed of the wind: w

    b.

	Rate	Time	Distance
Initial trip	$224 - w$	$6\frac{2}{3}$	$\frac{20}{3}(224 - w)$
Return trip	$224 + w$	5	$5(224 + w)$

    c.

    Initial trip $d = \dfrac{20}{3}(224 - w)$

    Return trip $d = 5(224 + w)$

    d. Length of the trip one way:
    $$\frac{20}{3}(224 - w) = 5(224 + w)$$

    e.
    $$\frac{20}{3}(224 - w) = 5(224 + w)$$
    $$3\left[\frac{20}{3}(224 - w)\right] = 3[5(224 + w)]$$
    $$20(224 - w) = 15(224 + w)$$
    $$4480 - 20w = 3360 + 15w$$
    $$\underline{\quad + 20w \qquad\qquad + 20w\quad}$$
    $$4480 = 3360 + 35w$$
    $$\underline{-3360 \quad -3360\quad}$$
    $$1120 = 35w$$
    $$\frac{1120}{35} = \frac{35w}{35}$$
    $$32 = w$$

    The speed of the wind was 32 mph.

15.
$$7 \times 98 = 7 \times (100 - 2)$$
$$= 7 \times 100 - 7 \times 2$$
$$= 700 - 14$$
$$= 686$$

17.
$$80 \times 99 = 80 \times (100 - 1)$$
$$= 80 \times 100 - 80 \times 1$$
$$= 8000 - 80$$
$$= 7920$$

19.

$$101 \times 50 = 50 \times (100 + 1)$$
$$= 50 \times 100 + 50 \times 1$$
$$= 5000 + 50$$
$$= 5050$$

21.

$$202 \times 45 = 45 \times (200 + 2)$$
$$= 45 \times 200 + 45 \times 2$$
$$= 9000 + 90$$
$$= 9090$$

23.

$$201 \times 36 = 36 \times (200 + 1)$$
$$= 36 \times 200 + 36 \times 1$$
$$= 7200 + 36$$
$$= 7236$$

Chapter 9 Review

1. False; The distributive law is used to multiply an outside factor by each term inside the parentheses.

3. True

5. False; The concentration of the mixture depends on the amount of each solution used.

7. False; $a(b+c) = ab + ac$

9. True

11.
$$3x - 5x - (-7x) = 3x - 5x + 7x$$
$$= 5x$$

13. $10 - 8(2s) = 10 - 16s$

15.
$$-6(7 - 5w) = -6(7) - (-6)(5w)$$
$$= -42 + 30w$$

17. $5 + (-3c - 5c) = 5 + (-8c) = 5 - 8c$

19.
$$-6a - 4(-6a - 4)$$
$$= -6a - 4(-6a) - 4(-4)$$
$$= -6a + 24a + 16$$
$$= 18a + 16$$

21.
$$-3(2 - 2N) + 4(6N - 5)$$
$$= -6 + 6N + 24N - 20$$
$$= 30N - 26$$

23.
$$3 - 2d - d = 0$$
$$3 - 3d = 0$$
$$\underline{-3 \qquad -3}$$
$$-3d = -3$$
$$\frac{-3d}{-3} = \frac{-3}{-3}$$
$$d = 1$$

25.
$$5r + 7(8 - r) = 12(4)$$
$$5r + 56 - 7r = 48$$
$$-2r + 56 = 48$$
$$\underline{-56 \quad -56}$$
$$-2r = -8$$
$$\frac{-2r}{-2} = \frac{-8}{-2}$$
$$r = 4$$

27.
$$4(d - 3) = -2(3d - 6)$$
$$4d - 12 = -6d + 12$$
$$\underline{+6d \qquad +6d}$$
$$10d - 12 = 12$$
$$\underline{+12 \ +12}$$
$$10d = 24$$
$$\frac{10d}{10} = \frac{24}{10}$$
$$d = \frac{24}{10} = \frac{12}{5}$$

29.
$$6u - 9(2u - 5) = 6 - 3(4 + 3u)$$
$$6u - 18u + 45 = 6 - 12 - 9u$$
$$-12u + 45 = -6 - 9u$$
$$\underline{+9u \qquad \qquad +9u}$$
$$-3u + 45 = -6$$
$$\underline{-45 \quad -45}$$
$$-3u = -51$$
$$\frac{-3u}{-3} = \frac{-51}{-3}$$
$$u = 17$$

31.   Number of men: x

$$x + 7x = 56$$
$$8x = 56$$
$$\frac{8x}{8} = \frac{56}{8}$$
$$x = 7$$

There are 7 men.

33.   Width of yard: w

$$2w + 2l = P$$
$$2w + 2(38) = 120$$
$$2w + 76 = 120$$
$$\underline{-76 \quad -76}$$
$$2w = 44$$
$$\frac{2w}{2} = \frac{44}{2}$$
$$w = 22$$

The width is 22 meters.

35.

$$\frac{-5}{6}t = \frac{-10}{3}$$
$$6\left(\frac{-5}{6}t\right) = \left(\frac{-10}{3}\right)6$$
$$-5t = -20$$
$$\frac{-5t}{-5} = \frac{-20}{-5}$$
$$t = 4$$

37.

$$\frac{2}{3}b + 1 = \frac{7}{3}$$
$$3\left(\frac{2}{3}b + 1\right) = \left(\frac{7}{3}\right)3$$
$$3\left(\frac{2}{3}b\right) + 3(1) = \left(\frac{7}{3}\right)3$$
$$2b + 3 = 7$$
$$2b = 4$$
$$b = 2$$

39.

$$\frac{1}{2}d + 5 = \frac{23}{4} - d$$
$$4\left(\frac{1}{2}d + 5\right) = \left(\frac{23}{4} - d\right)4$$
$$4\left(\frac{1}{2}d\right) + 4(5) = \left(\frac{23}{4}\right)4 + (-d)4$$
$$2d + 20 = 23 - 4d$$
$$6d + 20 = 23$$
$$6d = 3$$
$$d = \frac{1}{2}$$

41.   a. $0.08(20,000 - d)$ dollars

      b.

$$0.08(20,000 - d) = 0.08(20,000 - 3400)$$
$$= 0.08(16,600)$$
$$= \$1328$$

Aaron will pay $1328 in taxes.

      c.

$$1200 = 0.08(20,000 - d)$$
$$1200 = 1600 - 0.08d$$
$$\underline{-1600 \quad -1600}$$
$$-400 = -0.08d$$
$$\frac{-400}{-0.08} = \frac{-0.08d}{-0.08}$$
$$5000 = d$$

He must have $5000 in deductions.

43.   a. $\frac{2}{3}(R - 45,000)$ dollars

      b.

$$\frac{2}{3}(R - 45,000) = \frac{2}{3}(120,000 - 45,000)$$
$$= \frac{2}{3}(75,000)$$
$$= 50,000$$

Julie will receive $50,000.

      c.

$$75,000 = \frac{2}{3}(R - 45,000)$$
$$75,000 = \frac{2}{3}R - 30,000$$
$$\underline{+30,000 \qquad +30,000}$$
$$105,000 = \frac{2}{3}R$$
$$\frac{3}{2}(105,000) = \left(\frac{2}{3}R\right)\frac{3}{2}$$
$$157,500 = R$$

The revenue must be $157,500.

45.   a.

Investment ($) in savings	Investment ($) in money-market	Earnings ($) from savings	Earnings ($) from money-market	Total ($) earnings
100	1900	4	114	118
500	1500	20	90	110
1000	1000	40	60	100
1200	800	48	48	96
x	$2000 - x$	$0.04x$	$0.06(2000 - x)$	$0.04x + 0.06(2000 - x)$

b.

$$0.04x + 0.06(2000 - x) = 100.80$$
$$0.04x + 120 - 0.06x = 100.80$$
$$120 - 0.02x = 100.80$$
$$\underline{-120 \qquad\qquad -120}$$
$$-0.02x = -19.20$$
$$\frac{-0.02x}{-0.02} = \frac{-19.20}{-0.02}$$
$$x = 960$$

You invested $960 in savings and $2000 - 960 = \$1040$ in the money-market account.

47.   a.   Women; women

      b. $P = rW = (0.35)(120) = 42$ women in English
            $P = rW = (0.80)(60) = 48$ women in Math

      c. $120 + 60 = 180$ women candidates
            $42 + 48 = 90$ women admitted
            $r = \dfrac{P}{W} = \dfrac{90}{180} = 0.50$

50% of the women candidates were admitted to the university.

      d. $P = rW = (0.30)(80) = 24$ men in English
            $P = rW = (0.75)(160) = 120$ men in Math

      e. $80 + 160 = 240$ men candidates
            $24 + 120 = 144$ men admitted
            $r = \dfrac{P}{W} = \dfrac{144}{240} = 0.60$

60% of the men candidates were admitted to the university. This percentage is higher than the corresponding percentage for women even though the admission rate for women was higher in the individual departments.

49. a.

	Rate (miles/hour)	Time (hours)	Distance (miles)
Missile	600	t	$600t$
Bat-jet	720	$t - \dfrac{1}{4}$	$720\left(t - \dfrac{1}{4}\right)$

b.

$$600t = 720\left(t - \frac{1}{4}\right)$$
$$600t = 720t - 180$$
$$\underline{-720t \quad -720t}$$
$$-120t = -180$$
$$\frac{-120t}{-120} = \frac{-180}{-120}$$
$$t = 1.5$$

It will take the Bat-jet 1.5 hours to overtake the missile.

c. $d = 600t = 600(1.5) = 900$

Since the Bat-jet will overtake the missile in 900 miles, Batman will catch the missile before it destroys Metropolis.

51. $62 \times 50 = \dfrac{62}{2} \times 100 = 31 \times 100 = 3100$

53. $27 + 48 = (27 + 50) - 2 = (77) - 2 = 75$

55.
$$47 - 18 = 47 - (17 + 1)$$
$$= (47 - 17) - 1$$
$$= 30 - 1$$
$$= 29$$

57.
$$14 + 38 + 16 + 12 = (14 + 16) + (38 + 12)$$
$$= 30 + 50$$
$$= 80$$

59.
$$6 \times 299 = 6(300 - 1)$$
$$= 6 \times 300 - 6 \times 1$$
$$= 1800 - 6$$
$$= 1794$$

61. 45,678 is divisible by 2 since it is even.
45,678 is divisible by 3 since $4 + 5 + 6 + 7 + 8 = 30$ is divisible by 3.
45,678 is not divisible by 4 since 78 is not divisible by 4.
45,678 is not divisible by 5 since it does not end in 0 or 5.
45,678 is not divisible by 9 since $4 + 5 + 6 + 7 + 8 = 30$ is not divisible by 9.
45,678 is not divisible by 10 since it does not end in 0.
45,678 is divisible by 2 and 3.

Exercise A.1

1. $\dfrac{6}{10} = \dfrac{2 \cdot 3}{2 \cdot 5} = \dfrac{\cancel{2} \cdot 3}{\cancel{2} \cdot 5} = \dfrac{3}{5}$

3. $\dfrac{12}{75} = \dfrac{3 \cdot 4}{3 \cdot 25} = \dfrac{\cancel{3} \cdot 4}{\cancel{3} \cdot 25} = \dfrac{4}{25}$

5. $\dfrac{24}{72} = \dfrac{2 \cdot 2 \cdot 2 \cdot 3}{2 \cdot 2 \cdot 2 \cdot 3 \cdot 3} = \dfrac{\cancel{2} \cdot \cancel{2} \cdot \cancel{2} \cdot \cancel{3}}{\cancel{2} \cdot \cancel{2} \cdot \cancel{2} \cdot \cancel{3} \cdot 3} = \dfrac{1}{3}$

7. $\dfrac{36}{48} = \dfrac{2 \cdot 2 \cdot 3 \cdot 3}{2 \cdot 2 \cdot 2 \cdot 2 \cdot 3} = \dfrac{\cancel{2} \cdot \cancel{2} \cdot 3 \cdot 3}{\cancel{2} \cdot \cancel{2} \cdot 2 \cdot 2 \cdot \cancel{3}} = \dfrac{3}{4}$

9. $\dfrac{63}{105} = \dfrac{3 \cdot 3 \cdot 7}{3 \cdot 5 \cdot 7} = \dfrac{\cancel{3} \cdot 3 \cdot \cancel{7}}{\cancel{3} \cdot 5 \cdot \cancel{7}} = \dfrac{3}{5}$

11.

$$\dfrac{216}{18} = \dfrac{2 \cdot 2 \cdot 2 \cdot 3 \cdot 3 \cdot 3}{2 \cdot 3 \cdot 3}$$

$$= \dfrac{2 \cdot 2 \cdot 2 \cdot \cancel{3} \cdot \cancel{3} \cdot 3}{\cancel{2} \cdot \cancel{3} \cdot \cancel{3}}$$

$$= 12$$

13. $BF = 9 \div 3 = 3$

$\dfrac{2}{3} = \dfrac{2 \cdot BF}{3 \cdot BF} = \dfrac{2 \cdot 3}{3 \cdot 3} = \dfrac{6}{9}$

15. $BF = 8 \div 2 = 4$

$\dfrac{3}{2} = \dfrac{3 \cdot BF}{2 \cdot BF} = \dfrac{3 \cdot 4}{2 \cdot 4} = \dfrac{12}{8}$

17. $BF = 100 \div 5 = 20$

$\dfrac{2}{5} = \dfrac{2 \cdot BF}{5 \cdot BF} = \dfrac{2 \cdot 20}{5 \cdot 20} = \dfrac{40}{100}$

19. $BF = 72 \div 3 = 24$

$\dfrac{1}{3} = \dfrac{1 \cdot BF}{3 \cdot BF} = \dfrac{1 \cdot 24}{3 \cdot 24} = \dfrac{24}{72}$

21. $BF = 144 \div 8 = 18$

$\dfrac{5}{8} = \dfrac{5 \cdot BF}{8 \cdot BF} = \dfrac{5 \cdot 18}{8 \cdot 18} = \dfrac{90}{144}$

23. $BF = 6 \div 3 = 2$

$\dfrac{0}{3} = \dfrac{0 \cdot BF}{3 \cdot BF} = \dfrac{0 \cdot 2}{3 \cdot 2} = \dfrac{0}{6}$

Exercise A.2

1. $\dfrac{2}{3}\cdot\dfrac{5}{7}=\dfrac{2\cdot5}{3\cdot7}=\dfrac{10}{21}$

3. $\dfrac{6}{7}\cdot\dfrac{14}{15}=\dfrac{2\cdot3}{7}\cdot\dfrac{2\cdot7}{3\cdot5}=\dfrac{2\cdot2}{1\cdot5}=\dfrac{4}{5}$

5. $\dfrac{12}{16}\cdot\dfrac{18}{27}=\dfrac{3\cdot4}{2\cdot2\cdot4}\cdot\dfrac{2\cdot9}{3\cdot9}=\dfrac{1}{2}$

7. $\dfrac{28}{56}\cdot\dfrac{10}{15}=\dfrac{4\cdot7}{2\cdot4\cdot7}\cdot\dfrac{2\cdot5}{3\cdot5}=\dfrac{1}{3}$

9. $\dfrac{21}{48}\cdot\dfrac{88}{77}=\dfrac{3\cdot7}{2\cdot3\cdot8}\cdot\dfrac{8\cdot11}{7\cdot11}=\dfrac{1}{2}$

11. $\dfrac{24}{20}\cdot\dfrac{24}{36}\cdot\dfrac{3}{4}=\dfrac{4\cdot6}{4\cdot5}\cdot\dfrac{4\cdot6}{6\cdot6}\cdot\dfrac{3}{4}=\dfrac{3}{5}$

13.
$$\begin{aligned}
\frac{3}{4}\div\frac{5}{8}&=\frac{3}{4}\cdot\frac{8}{5}\\
&=\frac{3}{4}\cdot\frac{2\cdot4}{5}\\
&=\frac{3\cdot2}{5}\\
&=\frac{6}{5}
\end{aligned}$$

15.
$$\begin{aligned}
\frac{7}{3}\div\frac{28}{5}&=\frac{7}{3}\cdot\frac{5}{28}\\
&=\frac{7}{3}\cdot\frac{5}{4\cdot7}\\
&=\frac{5}{3\cdot4}\\
&=\frac{5}{12}
\end{aligned}$$

17.
$$\begin{aligned}
\frac{4}{5}\div6&=\frac{4}{5}\cdot\frac{1}{6}\\
&=\frac{2\cdot2}{5}\cdot\frac{1}{2\cdot3}\\
&=\frac{2}{5\cdot3}\\
&=\frac{2}{15}
\end{aligned}$$

19.
$$\begin{aligned}
4\div\frac{2}{9}&=\frac{4}{1}\cdot\frac{9}{2}\\
&=\frac{2\cdot2}{1}\cdot\frac{9}{2}\\
&=\frac{2\cdot9}{1}\\
&=18
\end{aligned}$$

21.
$$\begin{aligned}
\frac{11}{2}\div\frac{3}{4}&=\frac{11}{2}\cdot\frac{4}{3}\\
&=\frac{11}{2}\cdot\frac{2\cdot2}{3}\\
&=\frac{11\cdot2}{3}\\
&=\frac{22}{3}
\end{aligned}$$

23.
$$\begin{aligned}
\frac{30}{24}\div\frac{18}{72}&=\frac{30}{24}\cdot\frac{72}{18}\\
&=\frac{2\cdot3\cdot5}{3\cdot8}\cdot\frac{8\cdot9}{2\cdot9}\\
&=\frac{5}{1}\\
&=5
\end{aligned}$$

Exercise A.3

1. $\dfrac{1}{11}+\dfrac{3}{11}=\dfrac{1+3}{11}=\dfrac{4}{11}$

3. $\dfrac{14}{15}-\dfrac{7}{15}=\dfrac{14-7}{15}=\dfrac{7}{15}$

5. $\dfrac{1}{6}+\dfrac{3}{6}=\dfrac{1+3}{6}=\dfrac{4}{6}=\dfrac{2}{3}$

7. $\dfrac{9}{10}-\dfrac{4}{10}=\dfrac{9-4}{10}=\dfrac{5}{10}=\dfrac{1}{2}$

9. $\dfrac{1}{6}+\dfrac{2}{6}+\dfrac{3}{6}=\dfrac{1+2+3}{6}=\dfrac{6}{6}=1$

11. $\dfrac{1}{5}+\dfrac{3}{5}-\dfrac{2}{5}=\dfrac{1+3-2}{5}=\dfrac{2}{5}$

13. $\dfrac{7}{8}-\dfrac{3}{8}-\dfrac{2}{8}=\dfrac{7-3-2}{8}=\dfrac{2}{8}=\dfrac{1}{4}$

15. $\dfrac{19}{25}-\dfrac{8}{25}+\dfrac{4}{25}=\dfrac{19-8+4}{25}=\dfrac{15}{25}=\dfrac{3}{5}$

17. $6=\underline{2}\cdot\underline{3}$
$3=3$
$LCD=2\cdot3=6$

19. $4=\underline{2\cdot2}$
$6=2\cdot\underline{3}$
$LCD=2\cdot2\cdot3=12$

21. $3=\underline{3}$
$7=\underline{7}$
$LCD=3\cdot7=21$

23. $12=\underline{2\cdot2}\cdot3$
$30=2\cdot3\cdot\underline{5}$
$LCD=2\cdot2\cdot3\cdot5=60$

25. $2=\underline{2}$
$3=\underline{3}$
$5=\underline{5}$
$LCD=2\cdot3\cdot5=30$

27. $6=\underline{2}\cdot3$
$9=\underline{3\cdot3}$
$15=3\cdot\underline{5}$
$LCD=2\cdot3\cdot3\cdot5=90$

29. $9=\underline{3\cdot3}$
$4=\underline{2\cdot2}$
$12=2\cdot2\cdot3$
$LCD=2\cdot2\cdot3\cdot3=36$

31. $24=\underline{2\cdot2\cdot2}\cdot3$
$36=2\cdot2\cdot\underline{3\cdot3}$
$12=2\cdot2\cdot3$
$LCD=2\cdot2\cdot2\cdot3\cdot3=72$

33. $2=\underline{2};\ 3=\underline{3}$
$LCD=2\cdot3=6$
$\dfrac{1}{2}+\dfrac{1}{3}=\dfrac{1\cdot3}{2\cdot3}+\dfrac{1\cdot2}{3\cdot2}$
$\qquad=\dfrac{3}{6}+\dfrac{2}{6}$
$\qquad=\dfrac{5}{6}$

35. $4=\underline{2\cdot2};\ 3=\underline{3}$
$LCD=2\cdot2\cdot3=12$
$\dfrac{3}{4}-\dfrac{2}{3}=\dfrac{3\cdot3}{4\cdot3}-\dfrac{2\cdot4}{3\cdot4}$
$\qquad=\dfrac{9}{12}-\dfrac{8}{12}$
$\qquad=\dfrac{1}{12}$

37. $8 = \underline{2 \cdot 2} \cdot 2$; $12 = 2 \cdot 2 \cdot \underline{3}$

LCD $= 2 \cdot 2 \cdot 2 \cdot 3 = 24$

$$\frac{5}{8} + \frac{1}{12} = \frac{5 \cdot 3}{8 \cdot 3} + \frac{1 \cdot 2}{12 \cdot 2}$$

$$= \frac{15}{24} + \frac{2}{24}$$

$$= \frac{17}{24}$$

39. $15 = \underline{3} \cdot \underline{5}$; $10 = \underline{2} \cdot 5$

LCD $= 2 \cdot 3 \cdot 5 = 30$

$$\frac{14}{15} - \frac{7}{10} = \frac{14 \cdot 2}{15 \cdot 2} - \frac{7 \cdot 3}{10 \cdot 3}$$

$$= \frac{28}{30} - \frac{21}{30}$$

$$= \frac{7}{30}$$

41. $2 = \underline{2}$; $3 = \underline{3}$; $5 = \underline{5}$

LCD $= 2 \cdot 3 \cdot 5 = 30$

$$\frac{1}{2} + \frac{1}{3} + \frac{2}{5} = \frac{1 \cdot 15}{2 \cdot 15} + \frac{1 \cdot 10}{3 \cdot 10} + \frac{2 \cdot 6}{5 \cdot 6}$$

$$= \frac{15}{30} + \frac{10}{30} + \frac{12}{30}$$

$$= \frac{37}{30}$$

43. $20 = 2 \cdot 2 \cdot \underline{5}$; $8 = \underline{2 \cdot 2 \cdot 2}$; $12 = 2 \cdot 2 \cdot \underline{3}$

LCD $= 2 \cdot 2 \cdot 2 \cdot 3 \cdot 5 = 120$

$$\frac{13}{20} + \frac{3}{8} - \frac{5}{12} = \frac{13 \cdot 6}{20 \cdot 6} + \frac{3 \cdot 15}{8 \cdot 15} - \frac{5 \cdot 10}{12 \cdot 10}$$

$$= \frac{78}{120} + \frac{45}{120} - \frac{50}{120}$$

$$= \frac{73}{120}$$

45. $50 = 2 \cdot \underline{5 \cdot 5}$; $25 = 5 \cdot 5$; $4 = \underline{2 \cdot 2}$

LCD $= 2 \cdot 2 \cdot 5 \cdot 5 = 100$

$$\frac{49}{50} - \frac{8}{25} - \frac{1}{4} = \frac{49 \cdot 2}{50 \cdot 2} - \frac{8 \cdot 4}{25 \cdot 4} - \frac{1 \cdot 25}{4 \cdot 25}$$

$$= \frac{98}{100} - \frac{32}{100} - \frac{25}{100}$$

$$= \frac{41}{100}$$

47. $30 = 2 \cdot \underline{3} \cdot \underline{5}$; $40 = \underline{2 \cdot 2 \cdot 2} \cdot 5$; $70 = 2 \cdot 5 \cdot \underline{7}$

LCD $= 2 \cdot 2 \cdot 2 \cdot 3 \cdot 5 \cdot 7 = 840$

$$\frac{29}{30} - \frac{21}{40} + \frac{9}{70} = \frac{29 \cdot 28}{30 \cdot 28} - \frac{21 \cdot 21}{40 \cdot 21} + \frac{9 \cdot 12}{70 \cdot 12}$$

$$= \frac{812}{840} - \frac{441}{840} + \frac{108}{840}$$

$$= \frac{479}{840}$$

Exercise A.4

1.

$$3\frac{2}{3} = \frac{3}{1} + \frac{2}{3}$$
$$= \frac{3 \cdot 3}{1 \cdot 3} + \frac{2}{3}$$
$$= \frac{9}{3} + \frac{2}{3}$$
$$= \frac{11}{3}$$

3.

$$12\frac{1}{2} = \frac{12}{1} + \frac{1}{2}$$
$$= \frac{12 \cdot 2}{1 \cdot 2} + \frac{1}{2}$$
$$= \frac{24}{2} + \frac{1}{2}$$
$$= \frac{25}{2}$$

5.

$$20\frac{4}{5} = \frac{20}{1} + \frac{4}{5}$$
$$= \frac{20 \cdot 5}{1 \cdot 5} + \frac{4}{5}$$
$$= \frac{100}{5} + \frac{4}{5}$$
$$= \frac{104}{5}$$

7.

$$4\frac{21}{50} = \frac{4}{1} + \frac{21}{50}$$
$$= \frac{4 \cdot 50}{1 \cdot 50} + \frac{21}{50}$$
$$= \frac{200}{50} + \frac{21}{50}$$
$$= \frac{221}{50}$$

9.

$$3\frac{1}{4} = \frac{3}{1} + \frac{1}{4} = \frac{3 \cdot 4}{1 \cdot 4} + \frac{1}{4} = \frac{12}{4} + \frac{1}{4} = \frac{13}{4}$$
$$1\frac{5}{8} = \frac{1}{1} + \frac{5}{8} = \frac{1 \cdot 8}{1 \cdot 8} + \frac{5}{8} = \frac{8}{8} + \frac{5}{8} = \frac{13}{8}$$

$$\frac{13}{4} + \frac{13}{8} = \frac{13 \cdot 2}{4 \cdot 2} + \frac{13}{8}$$
$$= \frac{26}{8} + \frac{13}{8}$$
$$= \frac{39}{8}$$

11.

$$5\frac{2}{3} = \frac{5}{1} + \frac{2}{3} = \frac{5 \cdot 3}{1 \cdot 3} + \frac{2}{3} = \frac{15}{3} + \frac{2}{3} = \frac{17}{3}$$
$$6\frac{3}{4} = \frac{6}{1} + \frac{3}{4} = \frac{6 \cdot 4}{1 \cdot 4} + \frac{3}{4} = \frac{24}{4} + \frac{3}{4} = \frac{27}{4}$$

$$\frac{17}{3} + \frac{27}{4} = \frac{17 \cdot 4}{3 \cdot 4} + \frac{27 \cdot 3}{4 \cdot 3}$$
$$= \frac{68}{12} + \frac{81}{12}$$
$$= \frac{149}{12}$$

13.

$$9\frac{3}{8} = \frac{9}{1} + \frac{3}{8} = \frac{9\cdot 8}{1\cdot 8} + \frac{3}{8} = \frac{72}{8} + \frac{3}{8} = \frac{75}{8}$$

$$2\frac{1}{2} = \frac{2}{1} + \frac{1}{2} = \frac{2\cdot 2}{1\cdot 2} + \frac{1}{2} = \frac{4}{2} + \frac{1}{2} = \frac{5}{2}$$

$$\frac{75}{8} - \frac{5}{2} = \frac{75}{8} - \frac{5\cdot 4}{2\cdot 4}$$
$$= \frac{75}{8} - \frac{20}{8}$$
$$= \frac{55}{8}$$

15.

$$7\frac{3}{8} = \frac{7}{1} + \frac{3}{8} = \frac{7\cdot 8}{1\cdot 8} + \frac{3}{8} = \frac{56}{8} + \frac{3}{8} = \frac{59}{8}$$

$$1\frac{7}{12} = \frac{1}{1} + \frac{7}{12} = \frac{1\cdot 12}{1\cdot 12} + \frac{7}{12} = \frac{12}{12} + \frac{7}{12}$$
$$= \frac{19}{12}$$

$$\frac{59}{8} - \frac{19}{12} = \frac{59\cdot 3}{8\cdot 3} - \frac{19\cdot 2}{12\cdot 2}$$
$$= \frac{177}{24} - \frac{38}{24}$$
$$= \frac{139}{24}$$

17.

$$7\frac{1}{3} = \frac{7}{1} + \frac{1}{3} = \frac{7\cdot 3}{1\cdot 3} + \frac{1}{3} = \frac{21}{3} + \frac{1}{3} = \frac{22}{3}$$

$$2\frac{1}{4} = \frac{2}{1} + \frac{1}{4} = \frac{2\cdot 4}{1\cdot 4} + \frac{1}{4} = \frac{8}{4} + \frac{1}{4} = \frac{9}{4}$$

$$7\frac{1}{3} \cdot 2\frac{1}{4} = \frac{22}{3} \cdot \frac{9}{4}$$
$$= \frac{\not{2}\cdot 11}{\not{3}} \cdot \frac{\not{3}\cdot 3}{\not{2}\cdot 2}$$
$$= \frac{33}{2}$$

19.

$$3\frac{3}{7} = \frac{3}{1} + \frac{3}{7} = \frac{3\cdot 7}{1\cdot 7} + \frac{3}{7} = \frac{21}{7} + \frac{3}{7} = \frac{24}{7}$$

$$2\frac{1}{12} = \frac{2}{1} + \frac{1}{12} = \frac{2\cdot 12}{1\cdot 12} + \frac{1}{12} = \frac{24}{12} + \frac{1}{12}$$
$$= \frac{25}{12}$$

$$3\frac{3}{7} \cdot 2\frac{1}{12} = \frac{24}{7} \cdot \frac{25}{12}$$
$$= \frac{2\cdot 1\not{2}}{7} \cdot \frac{25}{1\not{2}}$$
$$= \frac{50}{7}$$

21.

$$2\frac{1}{3} = \frac{2}{1} + \frac{1}{3} = \frac{2\cdot 3}{1\cdot 3} + \frac{1}{3} = \frac{6}{3} + \frac{1}{3} = \frac{7}{3}$$

$$5\frac{3}{5} = \frac{5}{1} + \frac{3}{5} = \frac{5\cdot 5}{1\cdot 5} + \frac{3}{5} = \frac{25}{5} + \frac{3}{5} = \frac{28}{5}$$

$$2\frac{1}{3} \div 5\frac{3}{5} = \frac{7}{3} \div \frac{28}{5}$$
$$= \frac{7}{3} \cdot \frac{5}{28}$$
$$= \frac{\not{7}}{3} \cdot \frac{5}{4\cdot \not{7}}$$
$$= \frac{5}{12}$$

23.

$$3\frac{3}{4} = \frac{3}{1} + \frac{3}{4} = \frac{3\cdot 4}{1\cdot 4} + \frac{3}{4} = \frac{12}{4} + \frac{3}{4} = \frac{15}{4}$$

$$1\frac{7}{8} = \frac{1}{1} + \frac{7}{8} = \frac{1\cdot 8}{1\cdot 8} + \frac{7}{8} = \frac{8}{8} + \frac{7}{8} = \frac{15}{8}$$

$$3\frac{3}{4} \div 1\frac{7}{8} = \frac{15}{4} \div \frac{15}{8}$$
$$= \frac{15}{4} \cdot \frac{8}{15}$$
$$= \frac{15}{\not{4}} \cdot \frac{2\cdot \not{4}}{15}$$
$$= 2$$

25.

$$3\overline{)11} \atop 3$$

$$-\underline{9}$$

$$2$$

$$\frac{11}{3} = 3\frac{2}{3}$$

27.

$$8\overline{)43} \atop 5$$

$$-\underline{40}$$

$$3$$

$$\frac{43}{8} = 5\frac{3}{8}$$

29.

$$16\overline{)107} \atop 6$$

$$-\underline{96}$$

$$11$$

$$\frac{107}{16} = 6\frac{11}{16}$$

31.

$$32\overline{)317} \atop 9$$

$$-\underline{288}$$

$$29$$

$$\frac{317}{32} = 9\frac{29}{32}$$

Exercise A.5

1. $0.17 = \dfrac{17}{100}$

3. $0.07 = \dfrac{7}{100}$

5. $0.023 = \dfrac{23}{1000}$

7. $0.6 = \dfrac{6}{10} = \dfrac{2 \cdot 3}{2 \cdot 5} = \dfrac{3}{5}$

9. $0.26 = \dfrac{26}{100} = \dfrac{2 \cdot 13}{2 \cdot 50} = \dfrac{13}{50}$

11. $0.375 = \dfrac{375}{1000} = \dfrac{3 \cdot 5 \cdot 5 \cdot 5}{8 \cdot 5 \cdot 5 \cdot 5} = \dfrac{3}{8}$

13.
$$2.25 = 2\dfrac{25}{100}$$
$$= \dfrac{2 \cdot 100}{1 \cdot 100} + \dfrac{25}{100}$$
$$= \dfrac{225}{100}$$
$$= \dfrac{3 \cdot 3 \cdot 5 \cdot 5}{2 \cdot 2 \cdot 5 \cdot 5}$$
$$= \dfrac{9}{4}$$

15.
$$3.60 = 3\dfrac{6}{10}$$
$$= \dfrac{3 \cdot 10}{1 \cdot 10} + \dfrac{6}{10}$$
$$= \dfrac{36}{10}$$
$$= \dfrac{2 \cdot 2 \cdot 3 \cdot 3}{2 \cdot 5}$$
$$= \dfrac{18}{5}$$

17.
```
      .84
25)21.00
  -200
   100
  -100
     0
```
$\dfrac{21}{25} = 0.84$

19.
```
      .46
50)23.00
  -200
   300
  -300
     0
```
$\dfrac{23}{50} = 0.46$

21.
```
        .31
100)31.00
   -300
    100
   -100
      0
```
$\dfrac{31}{100} = 0.31$

23.
```
         .031
1000)31.000
    -3000
     1000
    -1000
        0
```
$\dfrac{31}{1000} = 0.031$

25.

$$16\overline{)5.0000} \; {.3125}$$

$$-\underline{48}$$

$$20$$

$$-\underline{16}$$

$$40$$

$$-\underline{32}$$

$$80$$

$$-\underline{80}$$

$$0$$

$$\frac{5}{16} = 0.3125$$

27.

$$8\overline{)3.000} \; {.375}$$

$$-\underline{24}$$

$$60$$

$$-\underline{56}$$

$$40$$

$$-\underline{40}$$

$$0$$

$$\frac{3}{8} = 0.375$$

29.

$$6\overline{)5.0000} \; {.8333}$$

$$-\underline{48}$$

$$20$$

$$-\underline{18}$$

$$20$$

$$-\underline{18}$$

$$20$$

$$-\underline{18}$$

$$2$$

$$\frac{5}{6} = 0.8333... = 0.8\overline{3}$$

31.

$$11\overline{)3.0000} \; {.2727}$$

$$-\underline{22}$$

$$80$$

$$-\underline{77}$$

$$30$$

$$-\underline{22}$$

$$80$$

$$-\underline{77}$$

$$3$$

$$\frac{3}{11} = 0.2727... = 0.\overline{27}$$

Exercise A.6

1.
   a. 10
   b. 14.8
   c. 14.77
   d. 14.774

3.
   a. 80
   b. 76.3
   c. 76.28
   d. 76.283

5.
   a. 170
   b. 169.9
   c. 169.90
   d. 169.899

7.
   a. 5550
   b. 5545.9
   c. 5545.91
   d. 5545.910

9.
   a. 700
   b. 701.0
   c. 700.96
   d. 700.960

11.
   a. 20
   b. 20.0
   c. 19.95
   d. 19.951

13.
   a. 1.9
   b. 1.91
   c. 1.907

15.
   a. 0.9
   b. 0.92
   c. 0.920

17.
   a. 0.1
   b. 0.10
   c. 0.099

19.
   a. 6.2
   b. 6.17
   c. 6.170

Exercise A.7

1.
$$1.46$$
$$+\,3.27$$
$$4.73$$

3.
$$3.64$$
$$+\,0.36$$
$$4.00$$

5.
$$0.26$$
$$+\,1.40$$
$$1.66$$

7.
$$13.00$$
$$+\ \ 0.26$$
$$13.26$$

9.
$$6.200$$
$$+\,0.027$$
$$6.227$$

11.
$$31.20$$
$$+\ \ 3.25$$
$$34.45$$

13.
$$12.63$$
$$-\ \ 9.16$$
$$3.47$$

15.
$$7.00$$
$$-\ \ 1.26$$
$$5.74$$

17.
$$6.02$$
$$-\ \ 0.95$$
$$5.07$$

19.
$$12.10$$
$$-\ \ 2.36$$
$$9.74$$

21.
$$438.40$$
$$-\ \ 76.25$$
$$362.15$$

23.
$$13.0000$$
$$-\ \ 0.0006$$
$$12.9994$$

25.
$$2.49$$
$$+\ 6.30$$
$$8.79$$
$$-\ 5.08$$
$$3.71$$

27.
$$573.100$$
$$-\ \ 28.960$$
$$544.140$$
$$-\ \ \ 2.094$$
$$542.046$$

29.
$$465.50$$
$$-\ 127.00$$
$$338.50$$
$$+\ 238.58$$
$$577.08$$

Exercise A.8

1.
```
      6.8
×    0.6
    4.08
```

3.
```
     0.32
×    0.4
   0.128
```

5.
```
      2.04
×    0.02
   0.0408
```

7.
```
      4.012
×    0.03
   0.12036
```

9.
```
       2.5
×    1.3
       75
      25
     3.25
```

11.
```
      4.32
×    2.4
     1728
      864
   10.368
```

13.
```
      4.12
×    0.42
      824
     1648
   1.7304
```

15.
```
     0.032
×    0.12
       64
       32
   0.00384
```

17.
```
        4.51
   7)31.57
   −  28
        35
   −   35
         07
   −     7
          0
```

19.
```
          4.21875
   16)67.50000
   −  64
        35
   −   32
         30
   −     16
          140
   −      128
           120
   −       112
            80
   −        80
             0
```

21.
```
        13.7
   05)68.5
   −   5
       18
   −  15
        35
   −   35
         0
```

23.
```
        15.3
   47)719.1
   −   47
        249
   −    235
         141
   −     141
           0
```

25.

```
        605
   38)22990
   -  228
       19
   -     0
      190
   -  190
        0
```

27.

```
       230.4
   19)4377.6
   -  38
      57
   -  57
      07
   -   0
      76
   -  76
       0
```

29.

```
         4.5454
   154)700.0000
   -   616
       840
   -   770
       700
   -   616
       840
   -   770
       700
   -   616
        84
```

31.

```
       800
   8)6400
   - 64
      0
```

Exercise A.9

1. $15\% = 0.15$

3. $0.4\% = 0.004$

5. $6.8\% = 0.068$

7. $119\% = 1.19$

9. $3\dfrac{1}{4}\% = 3.25\% = 0.0325$

11. $\dfrac{2}{5}\% = 0.4\% = 0.004$

13. $0.33 = 33\%$

15. $0.504 = 50.4\%$

17. $0.787 = 78.7\%$

19. $0.0201 = 2.01\%$

21. $0.008 = 0.8\%$

23. $5.5 = 550\%$

25.
$$\begin{aligned}
35\% &= \frac{35}{100} \\
&= \frac{\cancel{5}\cdot 7}{\cancel{5}\cdot 20} \\
&= \frac{7}{20}
\end{aligned}$$

27.
$$\begin{aligned}
125\% &= \frac{125}{100} \\
&= \frac{\cancel{25}\cdot 5}{\cancel{25}\cdot 4} \\
&= \frac{5}{4}
\end{aligned}$$

29.
$$\begin{aligned}
60\% &= \frac{60}{100} \\
&= \frac{\cancel{20}\cdot 3}{\cancel{20}\cdot 5} \\
&= \frac{3}{5}
\end{aligned}$$

31.
$$0.90\% = 0.009 = \frac{9}{1000}$$

33.
$$\begin{aligned}
37\frac{1}{2}\% &= 37.5\% \\
&= 0.375 \\
&= \frac{375}{1000} \\
&= \frac{3\cdot 125}{8\cdot 125} \\
&= \frac{3}{8}
\end{aligned}$$

35.
$$\begin{aligned}
33\frac{1}{3}\% &= \left(\frac{99}{3} + \frac{1}{3}\right)\% \\
&= \frac{100}{3}\% \\
&= \frac{100}{3} \div 100 \\
&= \frac{100}{3} \cdot \frac{1}{100} \\
&= \frac{100}{300} \\
&= \frac{1}{3}
\end{aligned}$$

37. $\dfrac{3}{4} = 3 \div 4 = 0.75 = 75\%$

39. $\dfrac{3}{8} = 3 \div 8 = 0.375 = 37.5\%$

41. $\dfrac{9}{9} = 9 \div 9 = 1 = 100\%$

43. $\dfrac{9}{4} = 9 \div 4 = 2.25 = 225\%$

45. $\dfrac{12}{5} = 12 \div 5 = 2.4 = 240\%$

47. $\dfrac{1}{250} = 1 \div 250 = 0.004 = 0.4\%$

Exercise A.10

1. $7 + 10 = 10 + \underline{7}$

3. $(3 + 6) + 9 = \underline{3} + (6 + 9)$

5. $36 \cdot 147 = \underline{147} \cdot 36$

7. $(17 \cdot 2) \cdot 5 = 17 \cdot (\underline{2} \cdot \underline{5})$

9. $(5 + 9) + 4 = (9 + \underline{5}) + 4$

11.
$$\begin{aligned} 47 + 28 + 3 &= (47 + 3) + 28 \\ &= 50 + 28 \\ &= 78 \end{aligned}$$

13.
$$\begin{aligned} 26 + 37 + 3 + 4 &= (26 + 4) + (37 + 3) \\ &= 30 + 40 \\ &= 70 \end{aligned}$$

15.
$$\begin{aligned} 2 \cdot 7 \cdot 5 &= (2 \cdot 5) \cdot 7 \\ &= 10 \cdot 7 \\ &= 70 \end{aligned}$$

17.
$$\begin{aligned} 50 \cdot 13 \cdot 2 &= (50 \cdot 2) \cdot 13 \\ &= 100 \cdot 13 \\ &= 1300 \end{aligned}$$

19.
$$\begin{aligned} 4 \cdot 6 \cdot 5 \cdot 5 &= (4 \cdot 5) \cdot (6 \cdot 5) \\ &= 20 \cdot 30 \\ &= 600 \end{aligned}$$